Brewing Yeast Fermentation Performance

Edited by

KATHERINE SMART
Oxford Brookes University
Oxford

**Blackwell
Science**

© 2000 by
Blackwell Science Ltd
Editorial Offices:
Osney Mead, Oxford OX2 0EL
25 John Street, London WC1N 2BS
23 Ainslie Place, Edinburgh EH3 6AJ
350 Main Street, Malden
  MA 02148 5018, USA
54 University Street, Carlton
  Victoria 3053, Australia
10, rue Casimir Delavigne
  75006 Paris, France

Other Editorial Offices:

Blackwell Wissenschafts-Verlag GmbH
Kurfürstendamm 57
10707 Berlin, Germany

Blackwell Science KK
MG Kodenmacho Building
7–10 Kodenmacho Nihombashi
Chuo-ku, Tokyo 104, Japan

The right of the Authors to be identified as the
Authors of this Work has been asserted in
accordance with the Copyright, Designs and
Patents Act 1988.

First published 2000

Set in 10/12 pt Times
by DP Photosetting, Aylesbury, Bucks
Printed and bound in Great Britain by
MPG Books Ltd, Bodmin, Cornwall

DISTRIBUTORS

Marston Book Services Ltd
PO Box 269
Abingdon
Oxon OX14 4YN
(*Orders:*   Tel: 01235 465500
           Fax: 01235 465555)

USA
  Blackwell Science, Inc.
  Commerce Place
  350 Main Street
  Malden, MA 02148 5018
  (*Orders:*   Tel: 800 759 6102
             781 388 8250
           Fax: 781 388 8255)

Canada
  Login Brothers Book Company
  324 Saulteaux Crescent
  Winnipeg, Manitoba R3J 3T2
  (*Orders:*   Tel: 204 837-2987
             Fax: 204 837-3116)

Australia
  Blackwell Science Pty Ltd
  54 University Street
  Carlton, Victoria 3053
  (*Orders:*   Tel: 03 9347 0300
             Fax: 03 9347 5001)

A catalogue record for this title is available from the
British Library

ISBN 0-632-05451-4

Library of Congress
Cataloging-in-Publication Data
is available

For further information on
Blackwell Science, visit our website:
www.blackwell-science.com

This volume is dedicated to the memory of Steve Whisker, Brewer, Morrells Brewery Limited. During the 1st Brewing Yeast Fermentation Performance Congress, Steve organised an enjoyable and informative technical visit to Morrells Brewery Limited for congress delegates. He will be fondly remembered and sorely missed.

Contributors

B. Axcell
South African Breweries, Research and Development Department, Sandton, 2146, PO Box 782178, South Africa

P.C. Austin
Department of Electrical and Electronic Engineering, University of Auckland, Privat Bag 92019, Auckland, New Zealand

D.S. Bendiak
Molson Breweries, Centre for Innovation, 33 Carlingview Drive, Etobicoke, Ontario, Canada M9W 5E4

G. Bihl
South African Breweries, Research and Development Department, Sandton, 2146, PO Box 782178, South Africa

C.A. Boulton
Bass Brewers Ltd, The Technical Centre, PO Box 12, Cross Street, Burton on Trent, Staffordshire DE14 1XH, UK

A. Cameron-Clarke
South African Breweries, Research and Development Department, Sandton, 2146, PO Box 782178, South Africa

V.J. Clutterbuck
Bass Brewers Ltd, The Technical Centre, PO Box 12, Cross Street, Burton on Trent, Staffordshire DE14 1XH, UK

S. Cunningham
Aber Instruments Ltd, Science Park, Aberystwyth, Ceredigion SY23 3AH, UK

G. Dinsdale
BIOSI (Microbiology), Cardiff University of Wales, PO Box 915, Cardiff CF10 3TL, UK

G. Dunne
Department of Biochemistry and Molecular Biology, University College London, London WC1E 6BT, UK

S.C.P. Durnin
Bass Brewers Ltd, The Technical Centre, PO Box 12, Cross Street, Burton on Trent, Staffordshire DE14 1XH, UK

J.R.M. Hammond
Brewing Research International, Lyttel Hall, Nutfield, Surrey RH1 4HY, UK

J.A. Hodgson
Scottish Courage Brewing Ltd, Technical Centre, Sugarhouse Close, 160 Canongate, Edinburgh EH8 8DD, UK

C. Holmes
Alaskan Brewing Company, Juneau, Alaska, USA

G. Hulse
South African Breweries, Research and Development Department, Sandton, 2146, PO Box 782178, South Africa

T. Imai
Kirin Brewery Co, Ltd, Research Laboratory for Brewing, 1-17-1 Namamugi, Tsurumi-ku, Yokohama 230-8628, Japan

A.I. Kennedy
Scottish Courage Brewing Ltd, Technical Centre, Sugarhouse Close, 160 Canongate, Edinburgh EH8 8DD, UK

B. Lancashire
Brewing Research International, Lyttel Hall, Nutfield, Surrey RH1 4HY, UK

A. Lentini
Brew/Tech-Carlton and United Breweries Limited, 1 Bouverie Street, Carlton, Victoria 3053, Australia

D. Lloyd
BIOSI (Microbiology), Cardiff University of Wales, PO Box 915, Cardiff CF10 3TL, UK

J. Londesborough
VTT Biotechnology and Food Research, PO Box 1500, VTT, F-02044, Finland

V. Martin
School of Biological and Molecular Sciences, Oxford Brookes University, Oxford OX3 0BP, UK

G. Morakile
University of the Orange Free State, Department of Microbiology and Biochemistry, South Africa

P. Piper
Department of Biochemistry and Molecular Biology, University College London, London WC1E 6BT, UK

C.D. Powell
School of Biological and Molecular Sciences, Oxford Brookes University, Oxford OX3 0BP, UK

D.E. Quain
Bass Brewers Ltd, The Technical Centre, PO Box 12, Cross Street, Burton on Trent, Staffordshire DE14 1XH, UK

W. Quilliam
South African Breweries, Research and Development Department, Sandton, 2146, PO Box 782178, South Africa

M. Reinman
VTT Biotechnology and Food Research, PO Box 1500, VTT, F-02044, Finland

M.R. Rhymes
School of Biological and Molecular Sciences, Oxford Brookes University, Oxford OX3 0BP, UK

D.L. Rodgers
School of Biological and Molecular Sciences, Oxford Brookes University, Oxford OX3 0BP, UK

P. Rodgers
Brew/Tech-Carlton and United Breweries Limited, 1 Bouverie Street, Carlton, Victoria 3053, Australia

R. Siddique
School of Biological and Molecular Sciences, Oxford Brookes University, Oxford OX3 0BP, UK

K.A. Smart
School of Biological and Molecular Sciences, Oxford Brookes University, Oxford OX3 0BP, UK

G.D. Smith
Division of Molecular and Life Sciences, School of Science and Engineering, University of Abertay Dundee, Bell Street, Dundee DD1 1HG, UK

G.G. Stewart
The International Centre for Brewing and Distilling, Heriot-Watt University, Riccarton, Edinburgh EH14 4AS, UK

B. Taidi
Scottish Courage Brewing Limited, Technical Centre, Sugarhouse Close, 160 Canongate, Edinburgh, EH8 8DD, UK

T. Takeuchi
Kirin Brewery Co, Ltd, Research Laboratory for Brewing, 1-17-1 Namamugi, Tsurumi-ku, Yokohama 230-8628, Japan

H.A. Teass
McNab Inc, Mount Vernon, New York, NY, USA

P.A. Thurston
Scottish Courage Brewing Ltd, Berkshire Brewery, Imperial Way, Reading, Berkshire RG2 0PN, UK

R. Todd
  Aber Instruments Ltd, Science Park, Aberystwyth, Ceredigion SY23 3AH, UK

G.M. Walker
  Division of Molecular and Life Sciences, School of Science and Engineering, University of Abertay Dundee, Bell Street, Dundee DD1 1HG, UK

O.S. Younis
  The International Centre for Brewing and Distilling, Heriot-Watt University, Riccarton, Edinburgh EH14 4AS, UK

D.C.K. Yuen
  Department of Electrical and Electronic Engineering, University of Auckland, Privat Bag 92019, Auckland, New Zealand

Contents

**4   Relationship between cell deterioration and plasma membrane ATPase of
Saccharomyces cerevisiae   27**
TOSHIHIKO TAKEUCHI and TAKEO IMAI

**5   Review of metabolic activity tests and their ability to predict fermentation
performance   34**
DIRK BENDIAK

Preface

KATHERINE SMART

Beer quality is strongly influenced by the biochemical performance of the yeast during fermentation. Many intrinsic and extrinsic factors may affect the rate and extent of progression of fermentation and the quality of the final product.

Fermentation performance may be defined as the capacity of a brewing yeast to consistently exhibit four key attributes: cell proliferation, utilisation of fermentable carbohydrates leading to ethanol production (attenuation), the aggregation and subsequent sedimentation of biomass at the end of fermentation (flocculation) and flavour development.

The physiological condition of brewing yeast influences fermentation performance, and therefore brewers require consistent yeast quality and quantity. Predicting the capacity of pitching yeast to perform well during fermentations is complex but remains of great importance. Many parameters may be examined for this purpose and these are comprehensively reviewed in this volume.

Measuring yeast quality inevitably involves the differentiation between live and dead cells. The life of a brewing yeast cell commences as a newly formed bud or daughter cell and, depending on the nutritional and physical constraints of the immediate environment, the cell will grow and divide to progress through its replicative lifespan until it senesces and finally dies. Yet the progression from bud to old age may be prematurely halted due to physiological stress.

The quality of brewing yeast reflects strain selection, propagation procedures and the handling during and between successive fermentations. Throughout these processes the yeast is subjected to a rapidly changing environment. Stresses may be generated by the yeast's own biochemical activity or by process requirements in the brewery. The relationship between yeast stress responses, fermentation performance and subsequent beer quality have not been elucidated. However, it is recognised that best practice yeast handling regimes are required to minimise the occurrence of stress and maximise yeast activity. Identifying stressed phenotypes rapidly is therefore an important aspect of yeast quality monitoring, and the current status of our understanding is reported and reviewed in this volume.

With such a wealth of yeast quality monitoring analyses available it would be supposed that predicting brewing yeast activity and therefore fermentation performance is a relatively simple task. Unfortunately, the very fact that each assessment technique is based on one parameter or one aspect of metabolic activity, means that their usefulness in predicting all four attributes contributing to fermentation performance is questionable. Therefore controlling the impact of stress on brewing biomass, predicting yeast activity and ensuring consistent fermentation performance through successive fermentations remain areas of active interest for the brewing industry.

Part 1 Measuring Yeast Quality

1 From Bright Field to Fluorescence and Confocal Microscopy

DAVID LLOYD and GWENDA DINSDALE

Abstract As in all fields of applied microbiology, microscopical observation of organisms involved in brewing technology provides essential information. Even the simplest microscope plays an indispensable role in process monitoring (for counting yeasts, assessment of vitality using the methylene blue reduction test, early diagnosis of gross bacterial contamination, etc). Fluorescence microscopy provides more sensitive and reliable methods for measuring vitality, and also viability. Rapid quantification of these characteristics with excellent precision can be obtained either by flow cytometry or by digital image analysis. Newly developed fluorophores extend flow cytometric measurements to enable studies of almost any cellular constituent of interest, and confocal laser scanning microscopy can be employed to validate fluorophore localisation within organisms. The most useful fluorophore for the evaluation of the quality of a yeast inoculum is the oxonol dye bis(1,3-dibutylbarbituric acid trimethine oxonol), $DiBAC_4(3)$, excluded from cells that have a significant trans-plasma membrane potential. Comparisons with other methods for assessment of vitality (vigorous metabolic activity) and viability (capacity for proliferation) indicate the usefulness of the oxonol fluorescence method.

1.1 Introduction

Yeast quality monitoring ideally requires a straightforward, sensitive and rapidly performed test that gives an instant answer. In the context of brewing technology there is a considerable literature on this subject; that before 1987 has been reviewed by Jones[1], who stressed the need to emphasise measurements by replication ability. The traditional techniques of counting viable cell numbers by plates or slide-count methods still provide information for yeast populations, although there may be some problems (of obtaining representative samples from sedimented organisms, of separating individuals from flocs prior to spreading, of over- or under-estimating plate inoculum size, etc.). Re-pitching of the population followed by monitoring of subsequent population increase[2] is the only true measure of yeast *viability*, but this is usually not feasible or indeed possible. Determination of bud-scar frequency per cell provides an indication of the 'age' of an individual yeast[3], but retrospective insights are not the most appropriate predictors of future division capacity.

Similar snags attend attempts to define the *vitality* of yeast inocula[4]. Perhaps the most sensible approach is to measure the vigour of glycolytic flux, most easily as CO_2 production rates[5,6], and there are convenient methods which are applicable to multiple samples tested in a specially designed multi-well plate. The most widely used method, the methylene blue reduction test (either in its traditional form or as a modified procedure which includes counterstaining with safranine O[1]) is a method based on the permeability and reduction capacity of the organism, with blue cells showing up as the non-viable ('dead') cohort. This test may give an approximate estimate of yeast quality, but is likely to be both strain-sensitive and operator-dependent. Other methods either involve time-consuming and technically exacting extraction procedures (e.g. for adenine nucleotides[2]). Other very useful measures of yeast cellular functional integrity (e.g. a.c. impedance, dielectric permittivity measurements[7-9]) require special equipment, and even though not outrageously costly, not available in every company.

Having explored several methods for the evaluation of the survival characteristics of cider yeast during the long (often more than 1 month) fermentations employed in that industry[2,3,10-13], we have discovered that an excellent even if indirect indicator of yeast viability is its capacity to exclude the anionic oxonol dye bis(1,3-dibutylbarbituric acid trimethine oxonol), $DiBAC_4(3)$. Whereas towards the end of the fermentation, by which time ethanol had increased to 11% (v/v) and the much more cytotoxic higher alcohols e.g. to as much as 260 ppm in the case of phenylethanol, microscopic examination revealed morphologically aberrant organisms with extensively plasmolysed cell contents, and most of these yeasts still excluded oxonol (Plate 1). Correlated with this was their ability to proliferate as shown by short lag phases, and normal growth rates and yields obtained on re-pitching (Fig. 1.1(a,b)). On the other hand, various criteria of vitality showed that glycolytic activities (CO_2 production rates, as monitored by membrane inlet mass spectrometry), acidification power, and adenylate charge values had all declined considerably from those determined in freshly inoculated cultures. Therefore, we concluded that the oxonol test, using fluorescence microscopy, is a useful indicator of the viability of the population.

(a)

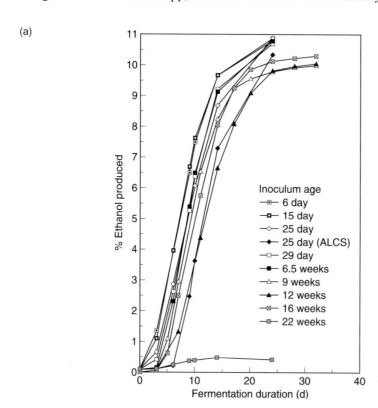

% Ethanol produced

Fermentation duration (d)

Inoculum age
—⊡— 6 day
—□— 15 day
—◇— 25 day
—◆— 25 day (ALCS)
—▱— 29 day
—■— 6.5 weeks
—△— 9 weeks
—▲— 12 weeks
—⊗— 16 weeks
—⊟— 22 weeks

(b)

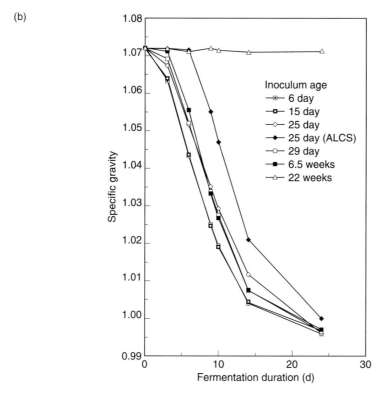

Fig. 1.1   Performance of cider yeast on re-pitching: (a) growth and (b) specific gravity.

However, it is more informative to have a statistically valid description of the population distribution of this property; because it interrogates individual cells at thousands per second, flow cytometry[14–17] provides an extremely rapid means of obtaining this. Differential counts of live (non-fluorescent) and dead (fluorescent) cells can easily be obtained. These counts provide an excellent indication of inoculum 'quality'.

1.2   Materials and methods

1.2.1   *Materials*

Yeast suspension was diluted with 20 mM Tris HCl buffer (pH 7.0) containing 2 mM $MgCl_2$ to a cell concentration of about 10^5 organisms per ml. $DiBAC_4(3)$ solution was prepared by dissolving 1 mg solid dye (Cat # B438 from Molecular Probes (www.probes.com) Europe BV PoortGebouw, Rijnsburgerweg 10, 2333 AA Leiden, The Netherlands (eurorder@probes.nl)) in 1 ml ethanol. This solution may be stored at $-18°C$ in the dark for 1 year at least. Dilute stock solution 10 µl to 1 ml with water (final concentration 10 µg/ml). This compound when dissolved in methanol has an absorption (excitation) maximum at 493 nm (molar extinction coeff. 123×10^3) and fluorescence emission maximum at 516 nm.

1.2.2 *Procedure*

(1) Incubate the yeast cell suspension with oxonol (final concentration 1 µg/ml) for 5 min. (2) Analyse using phase contrast (total cell count), and fluorescence microscopy, (for organisms with collapsed plasma membrane potential; blue-green fluorescence emission using blue excitation). Maintain observation while switching between bright field illumination and fluorescence (Plate 2) to reveal the relative proportion of viable and non-viable (fluorescent) organisms present in the sample. (3) If available, flow cytometry gives more rapid and statistically valid analyses. Use either a small air-cooled 3W Argon-ion laser for excitation at 488 nm, or a 100 W Hg-arc (lines at 405 nm and 435 nm selected using broad-band blue filters). Emission occurs at around 516 nm. Figure 1.2 shows flow cytometrically determined distributions of live and dead cells after oxonol staining. Positioning a cursor between the peaks gives a differential count. (4) Control experiments in which yeasts are heat-killed ($>65°C$ 15 min.) can be used to show that all organisms then become oxonol-permeable (Plate 2).

1.3 Discussion

Figure 1.3 shows the partitioning of the anionic oxonol ($pK = 4.2$) outside the yeast plasma membrane which has a trans-membrane potential (-ve inside) of about -60 mV. This potential is sustained by fluxes of H^+, Na^+, K^+, Ca^{2+} through channels and pumps.

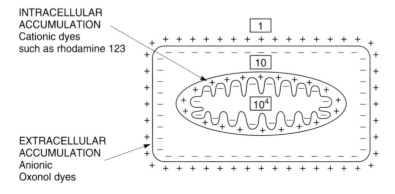

Fig. 1.3 Mechanism of oxonol exclusion in live cells.

Other voltage-sensitive dyes that have been used include Rhodamine 123[13,19], and various cyanine dyes[20–24] including JC-1[25]. These membrane potential sensitive dyes are all cationic and are accumulated only by live cells. Cells with active (energised) mitochondria (e.g. aerobically grown yeast electrophoretically accumulate these Nernstian dyes at ratios of $1:10:10^4$ from the extracellular medium, the cytosol and the mitochondrial matrix, respectively[10]. Therefore, these cationic dyes are useful for assessment of the activity of mitochondria (e.g. of an aerobically grown yeast inoculum (Plate 3(a)), or during the loss of mitochondrial function during anaerobic

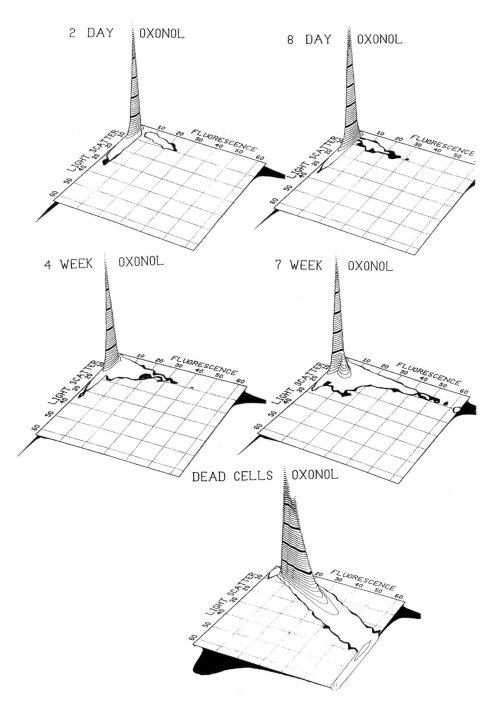

Fig. 1.2   Flow cytometry of oxonol-treated cider yeast at successive stages of the fermentation. Heat killed cells provide the control as an organism with zero plasma membrane potential that allows free penetration of the voltage sensitive dye.

fermentation (Plate 3(b)). In our hands Rhodamine 123 is the dye of choice for this. It is not useful as a viability indicator. Our results with cider yeast have been confirmed for baker's[26] and brewer's strains[27].

Other fluorescent dyes used for viability determination include propidium iodide[4,12], ChemChrome Y[26,28] and Fungolight[4,27]. Propidium iodide (red = dead) has often been employed with fluorescein diacetate (green = live) as a counterstain (Plate 4(a)), for yeasts or bacteria[29]. For cider yeast we have found that this method over-estimates dead cells, and other studies reveal discrepancies between results with propidium iodide and oxonol. The Fungolight kit (Molecular Probes, Plate 4(b)) provides excellent discrimination of live from dead cells using fluorescence microscopy: 'plasma membrane integrity and an (undefined) metabolic function are required to convert the yellow-green fluorescent intracellular staining into red-orange fluorescent intra-vacuolar structures' (Plate 5)[18]. However, it has not yet proved possible to devise a satisfactory flow cytometric protocol using this method[26].

ChemChrome Y (Chemmunex SA, Paris, France) is a fluorogen that is metabolized to give fluorescence only in culturable cells.[4]

Cell sorting provides conclusive evidence for the culturability of organisms[15,26].

1.4　Conclusions

DiBAC4(3) staining provides a convenient fluorescence viability assay for yeast. Counting of non-fluorescent individual cells gives a good correlation with the ability of these organisms to form colonies on yeast extract–peptone–glucose–agar medium or to grow in liquid cultures on repitching. Loss of ability to reduce methylene blue indicates diminished vitality rather than loss of viability. Other methods (CO_2 production rates, adenylate charge determinations, and acidification power) are also more useful as activity (vitality) indices rather than viability predictors.

Acknowledgements

Expert advice of Professor Basil Jarvis is acknowledged. M.G.D. held a BBSRC-CASE postgraduate studentship with H.P. Bulmer, Hereford.

References

(1)　Jones, R.P. (1987) Measures of yeast death I, II. *Process Biochemistry* **22/23**, 130–134.
(2)　Dinsdale, M.G., Lloyd, D., McIntyre, P. and Jarvis, B. (1999) Yeast vitality during cider fermentation: assessment by energy metabolism. *Yeast* **15**, 285–293.
(3)　Dinsdale, M.G. (1996) Yeast performance during ethanol accumulation in cider. Ph.D. thesis, University of Wales.
(4)　Lloyd, D. and Hayes, A.J. (1995) Vigour, vitality and viability of microorganisms. *FEMS Microbiology Letters* **133**, 1–7.
(5)　Lloyd, D., Kristensen, B. and Degn, H. (1983) Glycolysis and respiration in yeast: the Pasteur effect studied by mass spectrometry. *Biochemical Journal* **212**, 747–756.
(6)　Lloyd, D. and James C.J. (1987). The Pasteur effect in yeast: mass spectrometric monitoring of O_2 uptake and CO_2 and ethanol production. *FEMS Microbiology Letters* **42**, 27–31.

(7) Harris, C.M. and Kell, D.B. (1985) The estimation of microbial biomass. *Biosensors* 1, 17–84.
(8) Clarke, D.J., Blake-Coleman B.C., Carr R.J.G., Calder, M.R. and Atkinson, T. (1986) Monitoring reactor biomass. *Trends in Biotechnology* 4, 173–178.
(9) Kell, D.B. Markx, G.H., Davey, C.L. and Todd, R.W. (1990) Realtime measurement of cellular biomass: methods and applications. *Trends in Analytical Chemistry* 9, 190–194.
(10) Dinsdale, M.G., Lloyd, D. and Jarvis, B. (1995) Yeast vitality during cider fermentation; two approaches to the measurement of membrane potential. *Journal of the Institute of Brewing* 101, 453–458.
(11) Seward, R., Willetts, J.C., Dinsdale, M.G. and Lloyd, D. (1996) The effects of ethanol, hexan-1-ol and 2 phenylethanol on cider yeast growth, viability and energy status; synergistic inhibition. *Journal of the Institute of Brewing* 102, 439–443.
(12) Willetts, J.C., Seward, R., Dinsdale, M.G., Suller, M.T.E., Hill, B. and Lloyd, D. (1997) Vitality of cider yeast grown micro-aerobically with added ethanol, butan-1-ol or isobutanol. *Journal of the Institute of Brewing*, 103, 79–84.
(13) Lloyd, D., Moran, C.A., Suller, M.T.E., Dinsdale, M.G. and Hayes, A.J. (1996) Flow cytometric monitoring of rhodamine 123 and a cyanine dye uptake by yeast during cider fermentation. *Journal of the Institute of Brewing*, 102, 251–259.
(14) Lloyd, D. (1993). *Flow Cytometry in Microbiology*, Lloyd, D. (ed.), Springer-Verlag, Berlin.
(15) Davey, H.M. and Kell, D.B. (1996) Flow cytometry and cell sorting of heterogeneous microbial populations: the importance of sing-cell analyses. *Microbiological Reviews* 60, 641–696.
(16) Lloyd, D. (1999) Flow cytometry of yeasts. In *Current Protocols in Cytometry*, Robinson, J.P. (ed.), Wiley, New York.
(17) Davey, H.M., Weichart, D.H., Kell, D.B. and Kaprelyants, A.S. (1999) Estimating microbial viability using flow cytometry. In *Current Protocols in Cytometry*, Robinson, J.P. (ed.), Wiley, New York.
(18) Haugland, P.H. (1998) *Molecular Probes: Handbook of Fluorescent Probes and Research Chemicals*, 6th edn, Molecular Probes Inc., Eugene, OR 97402- 0414.
(19) Chen, L.B. (1988) Mitochondrial membrane potential in living cells. *Annual Review of Cell Biology* 4, 155–181.
(20) Gášková, D., Kurzweilová, H., Heman, P., Sigler, K., Plášek, J. and Mainský, J. (1994) Study of membrane potential changes of yeast cells caused by killer toxin K1. *Folia Microbiologia* 39, 505–560.
(21) Ordónez, J.V. and Wehman, N.M. (1995) Amphotericin suseptibility of *Candida* species assessed by rapid flow cytometric membrane potential assay. *Cytometry* 22, 154–157.
(22) Plášek, J. and Sigler, K. (1996) Slow fluorescent indicators of membrane potential: a survey of different approaches to probe response analysis. *Journal of Photochemistry and Photobiology* B 33, 101–124.
(23) Denksteinova, B., Sigler, K. and Plášek, J. (1996) Three fluorescent probes for the flow-cytometric assessment of membrane potential in *Saccharomyces cerevisiae*. *Folia Microbiologia* 41, 237–242.
(24) Gáščova, D., Brodská, B., Heman, P., Vee, J., Maínský, J., Sigler, K., Benada, Q. and Plášek, J. (1998) Fluorescent probing of membrane potential in walled cells: DiS-C_3(3) assay in *Saccharomyces cerevisiae*. *Yeast* 14, 1189–1197.
(25) Reers, M., Smith, T.W. and Chen, L.B. (1991) J-aggregate formation of a carboxycanine as a quantitative fluorescent indicator of membrane potential. *Biochemistry* 30, 4480–4486.
(26) Deere, D., Shen, J., Vesey, G., Bell, P., Bissinger, P. and Veal, D. (1998) Flow cytometry and cell sorting for yeast viability assessment and cell selection. *Yeast* 14, 147–160.
(27) Bell, P.J.L., Deere, D., Shen, J., Chapman, B., Bissinger, P.H., Attfield, P.V. and Veal, D.A. (1998) A flow cytometric method for rapid selecting novel industrial yeast hybrids. *Applied and Environmental Microbiology* 64, 1669–1672.
(28) Brailsford, M. and Gatley, S. (1993) Rapid analysis of microorganisms using flow cytometry. In *Flow Cytometry in Microbiology*, Lloyd, D. (ed.), Springer-Verlag, Berlin, pp. 171–180.
(29) Nebe-von Caron, G. and Badley R.A. (1996) Bacterial characterization by flow cytometry. In *Flow Cytometry Application in Cell Culture*, Al-Rubeai, M. and Emery, A.N. (eds), Marcel Dekker, New York, pp. 257–290.

2 Trehalose, Glycogen and Sterol

CHRIS BOULTON

Abstract Conventional modern brewery fermentation control regimes are largely devoted to ensuring that for any given beer quality, identical conditions are established at the completion of wort collection. These regimes assume that pitching yeast physiology is not a variable. However, inconsistencies can arise as a consequence of variability in the conditions encountered by pitching yeast during previous fermentations and in the periods of storage between cropping and re-pitching. Furthermore, it can be demonstrated that if corrective steps are not taken, these inconsistencies produce concomitant variability in fermentation performance and beer quality.

A current goal of brewing fermentation science is to develop procedures which are capable of rapidly assessing those facets of pitching yeast physiological condition which influence fermentation performance – the so-called 'yeast vitality' tests. Although such tests may be used simply to identify pitching yeast that is 'fit for purpose', it is much more desirable that they produce a result which can be used to predict the pitching rate and wort dissolved oxygen concentration which will produce optimum performance in the fermenter.

It has been suggested that pitching yeast condition can be gauged, to some extent, on the basis of macromolecular composition. Thus, the concentrations of certain cellular constituents may be indicative of the prior history of individual batches of yeast. In this regard, the concentrations of the carbohydrates, glycogen and trehalose and sterol lipids appear to be potentially useful metabolic markers. The significance of variations in the relative concentrations in pitching yeast of these metabolites is discussed, with particular reference to their usefulness both as predictors of fermentation performance and as indicators of the effectiveness of yeast husbandry in the brewery.

2.1 Introduction

A topic of continuing interest to brewing scientists is the development of procedures which are capable of assessing aspects of yeast physiology which influence fermentation performance – the so-called 'yeast vitality' tests[1]. Several of these tests are designed to probe specific aspects of pitching yeast metabolic activity, which might relate to subsequent behaviour in fermenter. However, it would also be predicted that the conditions to which pitching yeast had been exposed during previous fermentations and subsequent storage and handling would be reflected in changes in gross macromolecular composition. In particular, the intracellular concentrations of trehalose, glycogen and sterols are regarded as influential determinants of fermentation performance. This paper reviews the purported roles of these metabolites in brewing yeast and critically assesses their significance both as predictors of fermentation performance and indicators of the effectiveness of yeast husbandry in the brewery.

Free sterols are substituted tetracyclic alcohols and are important structural components of the yeast plasma-membrane, conferring fluidity, which is essential for proper membrane function[2]. Free sterols, ergosterol being the most prominent, account for up to 5% of the total lipid in yeast. Sterol esters, of which zymosterol is the most common, form up to 40% of the total yeast lipid[3]. The total sterol content of brewing yeast never exceeds 1 to 2% of the total dry cell weight. This contrasts with baker's yeast, cultivated under derepressing conditions, where the sterol pool typically accounts for up to 5% cell dry weight[4]. *De novo* synthesis of sterols derives from acetyl-CoA. The first part of the pathway culminates in the formation of squalene in a series of reactions which are anaerobic but require ATP and reducing power in the form of NADPH. Further metabolism of squalene involves an initial cyclisation reaction followed by complex series of molecular rearrangements, which yield ergosterol and other sterol intermediates. The precise sequence is strain specific; however, the initial cyclisation of squalene and some subsequent desaturation reactions require molecular oxygen. The quantity of oxygen supplied to wort at the start of fermentation controls the extent of subsequent sterol synthesis by yeast. Dilution of the pre-formed sterol pool as a consequence of cellular proliferation during the anaerobic phase of fermentation may be the factor, which limits overall growth during fermentation.

Glycogen is a high molecular weight polymer of α-D-glucose, which is synthesised from glucose 6-phosphate using uridine diphosphate as a carrier. Glycogen degradation yields glucose 1-phosphate. Glycogen accumulation occurs where growth is limited by a nutrient other than carbon[5]. It is generally accepted that yeast glycogen stores function as a reserve of carbon and energy which is utilised during periods of starvation or when the cells are undergoing adaptation to a new growth phase such as the diauxic shift[6]. In brewing yeast glycogen may account for up to 25% of the cell dry weight[7]. The glycogen reserve is used for maintenance energy during the period between the end of growth in fermenter and subsequent cropping and storage. Furthermore, it is reported that glycogen dissimilation provides energy and possibly carbon for sterol synthesis in the aerobic phase of fermentation, since membrane incompetence precludes the use of exogenous carbon[7].

Trehalose is a dimer of D-glucose (α-D-glucopyranosyl α-D-glucopyranoside) which like glycogen is also synthesised from glucose 6-phosphate and uses uridine diphosphate as the carrier for the second glucose molecule. Trehalose may also function as a reserve material; however, it seems more likely that it serves as a stress protectant. Thus, it has the ability to increase the thermotolerance of proteins and is capable of stabilising cellular membranes, thereby increasing the tolerance of cells to applied stresses such as desiccation, reduced water activity and elevated temperature[8]. The evidence for the specific stress protectant role for trehalose is in part suggested by the patterns of accumulation and dissimilation during batch growth, compared with changes in the glycogen pool[5]. In an aerobic batch culture of yeast, using a medium with a high sugar content, glycogen accumulation commences when growth begins to be limited by the disappearance of a nutrient other than carbon. The onset of significant trehalose accumulation is delayed until there is little or no exogenous supply of carbon. By inference some of the carbon used for trehalose synthesis may derive from glycogen dissimilation. Mobilisation of trehalose only occurs after prolonged starvation when the glycogen pool is relatively depleted. The disappearance of trehalose correlates with a rapid loss of viability.

Regulation of both glycogen and trehalose levels in yeast is complex and appears to be mediated, at least in part, by the cyclic-AMP-dependent phosphorylation/dephosphorylation metabolic cascades associated with global cellular signalling systems such as catabolite repression and de-repression[9]. However, under the conditions experienced by yeast in the brewing process, levels of trehalose are usually modest, typically less than 5% of the cell dry weight. Seemingly, trehalose concentrations are only elevated to the levels associated with bakers' yeast (10 to 20% cell dry weight) in the case of very high gravity brewing[10,11].

2.2   Materials and methods

2.2.1   Yeast samples

Yeast samples were obtained from brewery storage vessels.

2.2.2   Laboratory fermentations

Laboratory fermentations were performed in EBC tall-tubes.

2.2.3   Analyses

Sterols were extracted from dried yeast samples using petroleum ether after initial acid labilisation with 0.1M HCl and saponification with KOH. The extracted saponified sterols were determined as trimethylsilyl derivatives using capillary GC and cholestan-3β-ol as internal standard. Glycogen was determined using an enzymatic method[12]. Trehalose was determined using the anthrone method[13].

2.2.4   Yeast oxygenation

In some experiments yeast slurries were oxygenated as described previously[14].

2.3   Results and discussion

The data in Table 2.1 show the intracellular concentrations of trehalose, glycogen and total sterol in samples of pitching yeast, both ale and lager, removed from brewery storage vessels. The trehalose concentration was modest (mean 2.2% cell dry weight) in all cases with the exception of yeast derived from very high gravity fermentations. In the latter samples the mean trehalose concentration was 8.6% of the dry weight. These results are in accord with other published data[10,11], and apparently confirm that trehalose levels in pitching yeast are low under the conditions associated with brewing. Only very high gravity brewing elicits significant trehalose accumulation.

All the samples of yeast, apart from the last set, contained relatively high levels of glycogen (15 to 25% cell dry weight) and low levels (less than 0.2% cell dry weight) of total sterol. Although glycogen levels were relatively high there was considerable

Table 2.1 Trehalose, glycogen and total sterol content of samples of brewery pitching yeast.

Yeast sample	Glycogen (% dry wt)	Total sterol (% dry wt)	Trehalose (% dry wt)
Lager yeast 1 (Slurry, ex-14.5°P fermentation)	15.8	0.15	1.4
	14.7	0.17	2.2
	19.0	0.19	1.9
Lager yeast 2 (Slurry, ex-15°P fermentation)	22.5	0.09	2.9
	25.8	0.18	1.5
Lager yeast 2 (Slurry, ex-20°P fermentation)	18.4	0.22	8.4
	21.0	0.14	9.9
	24.8	0.12	7.6
Ale yeast 1 (Slurry, ex-10°P fermentation)	17.2	0.08	4.8
	14.9	0.16	2.8
Ale yeast 2 (Slurry, ex-10°P fermentation)	24.0	0.09	1.5
	23.8	0.12	1.0
	17.9	0.16	1.9
Ale yeast 2 (Pressed cake, ex-10°P fermentation)	5.4	0.25	2.2
	9.8	0.37	2.4
	7.3	0.24	1.7

variability in the absolute concentrations. Presumably this reflected variability in handling, in particular the duration of residence times in both fermenter and storage vessels. The last set of data relates to samples of an ale yeast stored as slabs of pressed cake in a cold store. In these samples the glycogen content was comparatively low (mean, 7.5% cell dry weight), whereas the sterol concentration was elevated (mean, 0.29% cell dry weight) compared with the same yeast stored as a slurry in beer. This is in accord with the assertion that exposure of pitching yeast to oxygen allows some sterol synthesis at the expense of glycogen dissimilation[15]. Thus, it would be predicted that the pressed yeast cake would be exposed to more oxygen during the interval between cropping and re-pitching compared with that stored as a slurry.

From a practical standpoint these data suggest that a measurement of glycogen content of pitching yeast alone would be a poor predictor of subsequent fermentation performance. Yeast stored under inappropriate conditions but with no exposure to oxygen should produce sub-optimal fermentation performance. However, pitching yeast exposed to oxygen during storage may also have a low glycogen content but elevated sterol. Such yeast should give a rapid fermentation when pitched into oxygenated wort. The poor correlation between glycogen content and fermentation performance was confirmed in the following experiment. Twelve samples of pitching yeast removed from brewery storage vessels were pitched, at the same viable rate, into wort in EBC tall-tubes. Fermentation performance was assessed as time to reach attenuation, and this was compared with the initial glycogen content of the pitching yeast. The correlation coefficient between these two parameters was 0.27.

The changes in the concentrations of trehalose, glycogen and sterols during fermentation are shown in Fig. 2.1. The sterol content, as a percentage of the cell dry weight, increased from 0.15% at pitch to approximately 1.2% after 20 h. The increase in sterol occurred during the aerobic phase and coincided with a rapid fall in glyco-

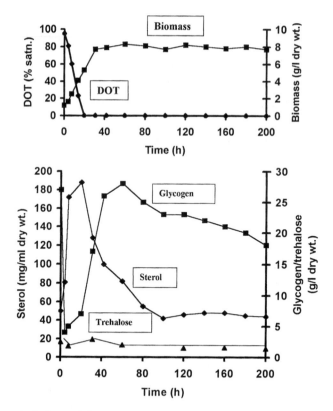

Fig. 2.1   Changes in the intracellular concentrations of glycogen, trehalose and total sterol compared with the total yeast biomass and the wort dissolved oxygen concentration during a 12° Plato lager fermentation.

gen. In the subsequent anaerobic phase of fermentation the sterol pool fell as the yeast biomass increased. The decrease in sterol concentration was accompanied by an increase in glycogen. In the final phase of the fermentation where yeast growth was completed sterol remained constant and low, whereas, glycogen gradually fell. Trehalose levels were low and remained relatively constant throughout fermentation.

These results support, at least superficially, the view that there is a causal relationship between utilisation of oxygen and glycogen, sterol synthesis and total yeast growth during fermentation. However, this is perhaps an over-simplistic picture. The data in Fig. 2.2 show some of the metabolic consequences of exposing pitching yeast to oxygen in the presence of exogenous maltose. It may be seen that maltose was utilised with no observable lag, and that this was accompanied by simultaneous dissimilation of glycogen. During the oxygenation process the exogenous ethanol concentration increased from approximately 18 g/l to more than 40 g/l. The magnitude of the increase in ethanol concentration was greater than could theoretically be accounted for from utilisation of the maltose alone. Therefore, this suggests that some of the ethanol derived from glycogen dissimilation. Obviously, in this experiment lack of membrane function did not preclude uptake of exogenous sugar.

In the experiment illustrated in Fig. 2.2, the intracellular squalene concentration at

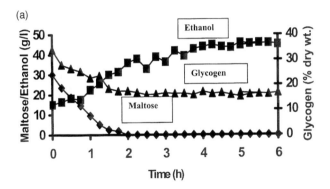

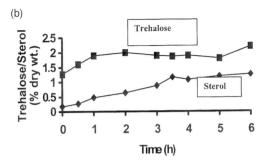

Fig. 2.2 Effect of oxygenating a lager pitching yeast slurry (35% wet wt/vol.) suspended in beer in the presence of maltose (3% wt/vol.). Yeast was oxygenated at 20°C, as described previously.

the start of the treatment was 1.4% of the cell dry weight. After 3 h oxygenation this decreased to 0.2% cell dry weight (results not shown). Thus, the decrease in the size of the squalene pool was sufficient to account for the total sterol synthesised. This suggests that little, if any, of the carbon released by glycogen dissimilation was utilised for *de novo* sterol synthesis.

Callaerts *et al.*[16] reported sterol synthesis at the expense of glycogen dissimilation when pitching yeast was exposed to oxygen under non-growing conditions. However, these authors also observed a simultaneous increase in intracellular trehalose concentration. They suggested that the trehalose accumulation may have been a consequence of the stresses associated with the oxygenation process, and that this would render the yeast more resistant to further applied stress. A similar experiment was performed here (Fig. 2.2(b)). The results were in accord with those of Callaerts *et al.*[16], in that there was a small increase in trehalose concentration from approximately 1.5 to 2.3% of the yeast dry weight. However, this did not confer an enhanced ability of the oxygenated yeast to withstand starvation during storage at an elevated temperature (18°C) compared with unoxygenated pitching yeast (Fig. 2.3). As may be seen, at this temperature the oxygenated yeast lost viability more rapidly than the unoxygenated control. Although not provable this suggests that glycogen, which would have been at a low level in the oxygenated yeast, is more important than

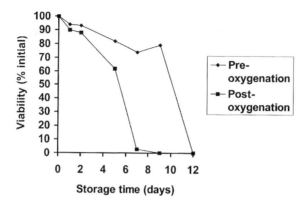

Fig. 2.3 Changes in the viability of yeast stored at 18°C, pre- and post-oxygenation.

trehalose in determining the ability of brewery pitching yeast to withstand periods of starvation.

The data presented in Table 2.1 and Fig. 2.2 possibly suggest that the brewing yeast strains used in this study are not capable of accumulating significant quantities of trehalose except in the case of very high gravity brewing. However, as indicated in Fig. 2.4, this is not the case. Here a lager brewing yeast, grown aerobically at 25°C on a semi-defined glucose-containing medium, was subjected to an abrupt heat shock by suddenly but transiently increasing the temperature to 45°C. This thermal stress was

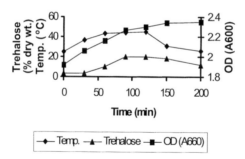

Fig. 2.4 Effect of heat shock on the trehalose concentration in lager yeast grown aerobically on semi-defined glucose-containing medium (yeast extract, peptone). Growth was assessed by measurement of optical density at 600 nm. The yeast culture was grown in a shake flask using an orbital incubator controlled at 25°C. At the start of the experiment the culture was transferred to an identical orbital incubator at a temperature of 50°C. After 2 h incubation the culture was returned to the 25°C incubator. Samples were removed for analysis at the times indicated.

accompanied by an immediate increase in the intracellular trehalose concentration from 2% to more than 15% of the cell dry weight. This result was similar to that reported for baker's yeast cultivated under aerobic conditions[17].

When pitching yeast is oxygenated under non-growing conditions the extent of sterol synthesis is similar to that which occurs in a conventional oxygenated wort fermentation (cf. Figs 2.1 and 2.2). It can also be demonstrated that within certain limits there is a direct correlation between the quantity of oxygen supplied to wort and the extent of subsequent yeast growth (Fig. 2.5). However, a combination of sterol-rich pitching yeast and wort oxygenation does not necessarily result in excessive yeast growth and rapid fermentation. For example, yeast removed from a propagator operated under fully aerobic conditions and then pitched into oxygenated wort produced slower than standard fermentations and a slightly reduced yeast crop[18]. This observation suggests that, although there is a direct relationship between the quantity of oxygen supplied to wort and the extent of yeast growth during fermentation, there is perhaps less evidence that this also correlates with the extent of sterol synthesis.

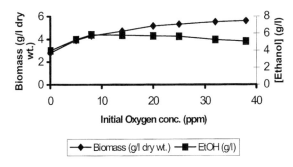

Fig. 2.5   Effect of varying the initial oxygen concentration on the yields of yeast and ethanol in 12°Plato lager fermentations. Fermentations were performed on a laboratory scale, as described previously[15].

2.4   Conclusions

The relatively low concentrations of trehalose detected and the lack of any modulation throughout the process suggest that this metabolite is of small significance to yeast under the conditions encountered during brewing. This was a surprising result considering the well-recognised phenomenon of trehalose accumulation as a stress response. Clearly, many stresses are imposed on yeast during the brewing cycle, including abrupt shifts in temperature, osmotic shocks, elevated hydrostatic pressure, oxidative stress, exposure to high concentrations of CO_2 and ethanol and periods of starvation. In the yeast strains examined here none of these stresses elicited trehalose accumulation except in the case of very high gravity brewing. Even then trehalose concentrations were less than those seen in baker's yeast or in the brewing yeast subjected to heat shock while growing aerobically. It may be concluded, therefore, that the presence of high trehalose in pitching yeast may be indicative of stressed yeast. However, it would also be likely that this stress would be evident from more conventional quality tests such as a simple methylene blue viability analysis.

Sterol synthesis usually occurs in the aerobic phase of fermentation in response to exposure to oxygen. Limited sterol synthesis may also occur if pitching yeast is exposed to oxygen in the interval between cropping and re-pitching. Under normal brewing conditions the quantity of sterol synthesised can be accounted for simply via cyclisation of the squalene pool. It follows, therefore, that there is little or no *de novo* synthesis of sterol using carbon derived from wort sugars or glycogen dissimilation. This is in accord with the observation that hydroxymethyl CoA reductase, the key enzyme in the pathway leading to the synthesis of squalene, is subject to glucose repression[19] and, therefore, should be inactive during the aerobic phase of fermentation. Coincidentally, this provides an explanation as to why derepressed baker's yeast apparently accumulates approximately fivefold more sterol than repressed brewer's yeast.

A correlation exists between the quantity of oxygen supplied to wort and the extent of yeast growth in fermentation. However, there is no similar correlation between yeast growth and sterol concentration. This is perhaps unsurprising in view of the fact that it is reported that only 10% of wort oxygen is used for sterol synthesis, 15% for unsaturated fatty acid synthesis, the remaining 75% having no known role[20]. It follows that analysis of the sterol content of pitching yeast would be of little predictive value.

Sterol synthesis from squalene requires energy and reducing power in the form of NADPH. This may be provided by glycogen dissimilation; however, the evidence provided here indicates that exogenous sugars would be used with equal facility in the absence of adequate glycogen reserves. Undoubtedly glycogen is mobilised immediately after pitching; however, it seems that exposure to oxygen is the metabolic trigger for this and there may be no direct linkage to sterolagenesis. Possibly some of the carbon is utilised for unsaturated fatty acid synthesis. Glycogen does seem to be of most importance for use as an energy reserve during yeast storage. The very high levels of glycogen observed in pitching yeast may have no more metabolic significance than as a reflection of the high carbon contents of most brewery worts. A similar argument could account for the elevated trehalose content of yeast used in very high gravity brewing.

Acknowledgements

The author thanks the Directors of Bass Brewers for permission to publish this paper and acknowledges the contributions made by Wendy Box, David Quain, Andrew Jones and Sean Durnin.

References

(1) Lentini, A. (1993) A review of the various methods available for monitoring the physiological status of yeast: viability and vitality. *Ferment* **6**, 321–327.
(2) Weete, J.D. (1989) Structure and function of sterols in fungi. *Advances in Lipid Research* **23**, 115–167.
(3) Rattray, J.B.M. (1988) Yeasts. In *Microbial Lipids*, Vol. 1, Ratledge, C. and Wilkinson, S.G. (eds), Academic Press, London, pp. 555–697.
(4) Quain, D.E. and Tubb, R.S. (1982) The importance of glycogen in brewing yeasts. *Technical Quarterly of the Master Brewers of the Americas* **19**, 29–33.
(5) Lillie, S.H. and Pringle, J.R. (1980) Reserve carbohydrate metabolism in *S. cerevisiae*: responses to nutrient limitation. *Journal of Bacteriology* **143**, 1384–1394.
(6) Farkas, V. (1989) Polysaccharide metabolism. In *The Yeasts*, Vol. 3, Rose, A.H. and Harrison, J.S. (eds), Academic Press, London, pp. 317–366.

(7) Quain, D.E., Thurston, P.A. and Tubb, R.S. (1981) The structural and storage carbohydrates of *S. cerevisiae*: changes during the fermentation of wort and a role for glycogen catabolism in lipid biosynthesis. *Journal of the Institute of Brewing* **87**, 108–111.

(8) Wiemken, A. (1990) Trehalose in yeast: stress protectant rather than reserve carbohydrate. *Antonie van Leeuwenhoek* **58**, 209–217.

(9) Panek, A.D. (1991) Storage carbohydrates. In *The Yeasts*, Vol. 4, Rose, A.H. and Harrison, J.S. (eds), Academic Press, London, pp. 655–678.

(10) Majara, M., O'Connor-Cox, E.S.C. and Axcell, B.C. (1996) Trehalose – an osmoprotectant and stress indicator compound in high and very high gravity brewing. *Journal of the American Society of Brewing Chemists* **54**, 149–154.

(11) Majara, M., O'Connor-Cox, E.S.C. and Axcell, B.C. (1996) Trehalose – a stress protectant and stress indicator compound for yeast exposed to adverse conditions. *Journal of the American Society of Brewing Chemists* **54**, 221–227.

(12) Quain, D.E. (1981) The determination of glycogen in yeasts. *Journal of the Institute of Brewing* **95**, 315–323.

(13) Herbert, D., Phipps, P.J. and Strange, R.E. (1971) Chemical analysis of microbial cells. *Methods in Microbiology* B **5**, 209–344.

(14) Boulton, C.A., Jones, A.R. and Hinchliffe, E. (1991) Yeast physiological condition and fermentation performance. *Proceedings of the European Brewing Congress* **23**, 385–392.

(15) Boulton, C.A. and Quain, D.E. (1987) Yeast, oxygen and the control of brewery fermentations. *Proceedings of the European Brewing Congress* **21**, 401–408.

(16) Callaerts, G., Iserentant, D. and Varachtert, H. (1993) Relationship between trehalose and sterol accumulation during oxygenation of cropped yeast. *Journal of the American Society of Brewing Chemists* **51**, 75–77.

(17) Hottiger, T., Boller, T. and Wiemken, A. (1989) Correlation of trehalose content and heat resistance in yeast mutants altered in the RAS/adenylate cyclase pathway: is trehalose a heat protectant? *FEBS Letters* **255**, 431–434.

(18) Boulton, C.A. and Quain, D.E. (2000). A novel system for the propagation of brewing yeast. *Proceedings of the European Brewing Congress* **27**, in press.

(19) Quain, D.E. and Haslam, J. (1979) The effects of catabolite derepression on the accumulation of steryl esters and the activity of 3-hydroxymethylglutaryl-CoA reductase in *S. cerevisiae*. *Journal of General Microbiology* **111**, 343–351.

(20) Kirsop, B.H. (1982) Developments in beer fermentation. *Topics in Enzyme and Fermentation Biotechnology* **6**, 79–131.

3 Rapid Mobilization of Intracellular Trehalose by Fermentable Sugars: a Comparison of Different Strains

MIKKO REINMAN and JOHN LONDESBOROUGH

Abstract Glucose- (or carbon catabolite-) induced signalling in laboratory and industrial yeasts was investigated by adding starved yeast to wort or artificial media containing sugars and measuring trehalase and trehalose. In artificial media, four lager yeasts and one ale yeast rapidly depleted their trehalose in response to either glucose or (more slowly) maltose. One lager strain mobilised trehalose slowly in response to glucose and not at all to maltose, but rapidly in wort. Maltose did not mobilise trehalose in industrial bakers' yeast nor in a maltose-negative laboratory strain (RH144). Maltose-induced mobilization of trehalose was lost when a lager yeast was grown in an artificial medium with glucose as the carbon source. The levels of glycogen, glycerol, trehalose and trehalose 6-phosphate synthase and the activation state of trehalase were followed through high gravity fermentations. Three brewer's strains rapidly depleted trehalose even when pitched into VHG (24°Plato) wort, despite its high osmolarity. These results suggest that brewer's strains have evolved carbon catabolite-induced signalling that responds to maltose. Which sugars trigger the signalling depends on how the yeast was grown. There are differences between strains in how they respond to the opposed effects of sugar availability and osmotic stress in VHG worts.

3.1 Introduction

The fermentation performance of cropped yeast seems to be better when it contains more glycogen[1] and trehalose[2]. When starved or otherwise derepressed yeast is added to readily fermentable sugars, several parallel regulatory processes are triggered that cause changes in enzyme activities and gene expression[3]. This occurs under laboratory conditions and when brewer's yeast is added to wort. One of the regulatory pathways works via a very rapid rise in cyclic AMP ('glucose-induced cAMP signal') and causes activation of trehalase and mobilization of trehalose[4]. Intensive investigations using genetically defined laboratory strains and glucose have not identified the initial sugar receptor(s). Fructose and sucrose can replace glucose, but little work has been reported with maltose. Apparently maltose is not able to cause a cAMP signal even in yeast containing constitutive maltose permease and maltase[5]. Maltose permease and maltase are themselves repressed, and the permease is inactivated when glucose is added to yeast, although the signal pathway involved is not known.

The glucose/maltose ratio in worts might affect the response of pitched yeast. Furthermore, in VHG worts yeast receives opposed signals: glucose-induced signalling mobilises trehalose whereas hyperosmotic stress increases trehalose (and glycerol). We measured changes in trehalose, glycogen and glycerol levels in several yeast strains in defined model systems and tall-tube fermentations. To our surprise, we found that maltose itself can trigger trehalose mobilization in brewer's strains.

3.2   Materials and methods

3.2.1   *Yeast*

Strains from VTT's collection were the lager yeasts, A15, A24, A64, A72 and A143, an ale yeast A60 and a laboratory strain of *S. cerevisiae*, RH144. Commercial bakers' yeast was bought in a shop. Except where mentioned otherwise, brewers' strains were grown in 16°Plato wort to stationary phase at 25°C in shake flasks. RH144 was grown in YPD (10 g yeast extract, 20 g peptone and 20 g glucose/l) in the same way.

3.2.2   *Brewing fermentations*

Fermentations were in steel tubes containing 2 l wort at 10°C or 20°C. 16°Plato wort (from a local brewery) was pitched with 5 g yeast/l and 24°Plato wort (from our pilot brewery) with 8 g yeast/l. Samples (30 to 100 ml) were taken anaerobically through membrane ports and centrifuged. The supernatants were used for sugar composition and gravity determinations. The yeast pellets were washed with cold 25 mM K phosphate pH 7.0/1 mM EDTA and suspended in this buffer to 300 mg/ml.

3.2.3   *Model pitching experiments*

Yeast slurry (50% in water) was added to wort or an artificial medium containing glucose, fructose or maltose (20 g/l) to a final concentration of 30 mg fresh yeast/ml at 10°C. At intervals 50 ml portions were centrifuged. The yeast pellets were washed and suspended as described above.

3.2.4   *Trehalose, glycerol, glycogen and enzyme assays*

Portions (0.5 ml, 150 mg fresh wt) of the washed yeast samples were boiled for 5 min, centrifuged and the precipitates washed. Trehalose[6] and glycerol were determined enzymatically from the supernatants. Portions (0.2 ml) were centrifuged and the yeast pellets frozen. Glycogen was extracted from them and assayed according to Schulze *et al.*[6]. Portions (1 to 2 ml) were centrifuged and the yeast pellets frozen. The thawed yeast was broken by shaking with glass beads in 50 mM HEPES/KOH pH 7.0/2 mM $MgCl_2$/1 mM EDTA/0.1 mM dithiothreitol containing protease inhibitors. The extracts were assayed for trehalose 6-phosphate synthase[7] (Tre6P synthase) and trehalase[8].

3.2.5   *Interpretation of trehalase assays*

The activation degree of metal-dependent, neutral trehalase was estimated by trehalase assays at 2 mM $CaCl_2$ (total trehalase) and 2 mM $MgCl_2$/1 mM EGTA (giving 1 mM Mg^{2+} and very low Ca^{2+}; activated trehalase). Results were corrected for the metal-independent acidic trehalase by subtracting the activity at 2 mM EDTA. According to Londesborough and Varimo[8], the relative activities of *in vitro* activated and non-activated trehalase are:

	$20\,\mu M\ Ca^{2+}$	$2\,mM\ Ca^{2+}$	$1\,mM\ Mg^{2+}$
Activated (phosphorylated *in vitro*)	45	95	40
Non-activated	3	50	0

The Mg/Ca activity ratio should be 0.42 for phosphorylated trehalase and zero for non-phosphorylated. Ratios up to 0.58 were observed, either because the *in vitro* phosphorylated enzyme[8] was not completely phosphorylated or because small amounts of free Ca^{2+} were still present in our Mg/EGTA assays.

3.2.6   *Maltose permease*

Maltose permease was determined essentially as already described[9] with 20 s incubations in 5 mM [^{14}C]maltose at 20°C.

3.3   Results and discussion

3.3.1   *Tall-tube fermentations*

Fermentations with A24 yeast are shown in Fig. 3.1. At 24°Plato and 10°C the fermentation abruptly slowed after reaching 15°Plato (Fig. 3.1(b)) but went to completion at 20°C (Fig. 3.1(c)). On pitching into 16°Plato wort, trehalose dropped by ≥90% in 4 h and glycogen dropped more slowly (Fig. 3.1(a)), in agreement with previous results[1,10]. Despite the high osmolarity, trehalose also dropped rapidly when this strain was pitched into 24°Plato wort (Figs 3.2 and 3.3). Similar results were obtained with two other lager yeasts (A15 and A143; not shown). By contrast, trehalose in a S. African lager strain is reported to increase on pitching into 25°Plato wort[10]. Strain variation is not surprising, since in this situation two regulatory systems are opposed, trehalose mobilization and accumulation in response to sugar availability and osmotic stress, respectively.

At 24°Plato, intracellular glycerol rose during the first day (Fig. 3.1(a,b)). This is the expected response to a high osmotic gradient across the cell membrane, which contains osmoreceptors (Sln1 and Sho1) that lead to increased glycerol synthesis and retention[11].

At 16°Plato and 10°C Tre6P synthase was present at a low level (2.3 ± 0.5 U/g yeast; the max. level is about 15 U/g) at least between 40 h and 140 h (Fig. 3.1(a)). At both 16°Plato and 24°Plato trehalose started to rise again after about 70 h at 10°C (Figs 3.1 and 3.2). This rise was accompanied by a marked decrease in the trehalase Mg/Ca ratio (Fig. 3.1(a)). Thus, this trehalose accumulation is caused by deactivation of trehalase in the presence of a rather constant low level of Tre6P synthase. Glycogen increased simultaneously, but glycerol *fell*, suggesting that the osmotic stress is decreasing. Conversion of hexose to two molecules of ethanol lowers water activity but relieves the trans-membrane osmotic gradient, because the membrane is permeable to ethanol. Yeast growth halted after the first 20 h and a second growth phase coincided with the rise in trehalose and glycogen (Fig. 3.1(a,b)), which caused 60% of the increase in dry mass. Trehalose reached higher levels (120 mg/g dry wt) at 24°Plato

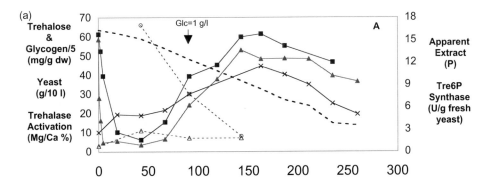

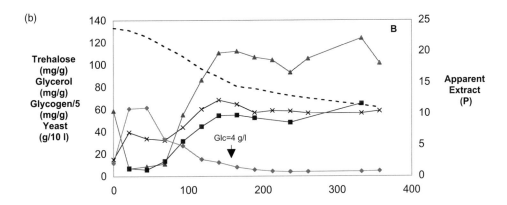

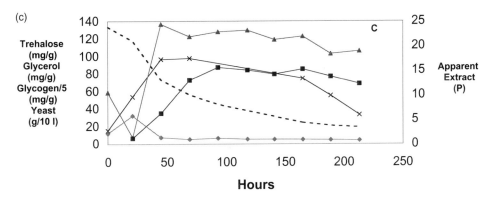

Fig. 3.1   Tall-tube fermentations with A24 yeast and (a) 16°Plato wort at 10°C; (b) 24°Plato wort at 10°C and (c) 24°Plato at 20°C. The apparent extract (- - - -), dry yeast concentration ($\times$), intracellular trehalose (▲), glycogen (■), and glycerol (◆), Tre6P synthase activity (- -△- -) and trehalase activation (- -○- -) are shown. The arrows in (a) and (b) indicate when glucose had dropped to 1 g/l and 4 g/l and 4 g/l, respectively.

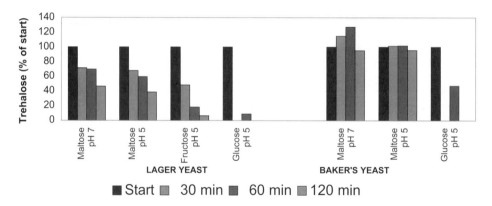

Fig. 3.2 Effect of different sugars on trehalose mobilisation in commercial brewer's and baker's yeasts. A24 lager yeast (54 mg trehalose/g dry mass; grown on wort) and commercial baker's yeast (140 mg trehalose/g dry mass; grown on molasses) were added at 10°C to YP medium, pH 5 or 7, containing maltose, fructose or glucose (20 g/l).

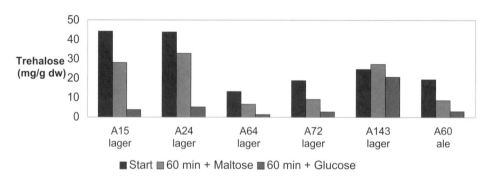

Fig. 3.3 Trehalose mobilisation in six brewing strains by maltose or glucose in YP medium, pH 7.

than at 16°Plato (50 mg/g dry wt) but was rather independent of temperature, whereas glycogen reached higher levels (400 mg/g dry wt) at 20°C than at 10°C (300 mg/g dry wt) but was rather independent of wort strength (Fig. 3.1).

3.3.2 Model pitching experiments

With A24 yeast, fructose mobilised trehalose almost as effectively as glucose, and maltose also caused mobilisation, although more slowly. By contrast, commercial bakers' yeast did not respond to maltose (Fig. 3.2). In a 16°Plato wort (21 g glucose/l, 4 g fructose/l, 78 g maltose/l and 21 g maltotriose/l) at 10°C trehalose was mobilised in A24 yeast almost as rapidly as in YP/2% glucose, and trehalase activation was marked in 30 min and complete in 2 h (Mg/Ca activity ratios were 0.02, 0.19 and 0.56, respectively, at 0, 30 and 120 min).

Maltose mobilised trehalose in four lager strains and one ale strain, but neither maltose nor glucose mobilised trehalose in lager strain A143 (Fig. 3.3). Nevertheless,

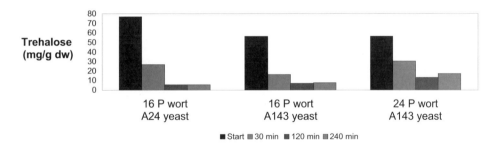

Fig. 3.4 Trehalose mobilisation in A24 and A143 lager yeasts in worts. Yeasts grown in 16°Plato wort were pitched into 16 and 24°Plato worts in tall-tubes.

trehalose was efficiently mobilised when this strain was pitched into either 16°Plato or 24°Plato worts (Fig. 3.4).

RH144 is a presumably maltose-negative laboratory strain that grows very poorly on YP/maltose. After growth on YP/2% glucose, its trehalose was mobilised by glucose, but not by maltose (Fig. 3.5) and no maltopermease was detected. After 4 days growth on either YP/2% glucose or YP/2% maltose, also A24 lager yeast lost its ability to respond to maltose by mobilising trehalose (Fig. 3.5: the increase in trehalose caused by maltose may represent uptake of trehalose [0.5 mg/ml] in the YP medium). These glucose-grown and maltose-grown A24 yeasts contained less maltose permease (65 µmol/h per g dry wt and 35 µmol/h per g dry wt) than did the wort-grown cells (200 µmol/h per g dry wt) in which maltose mobilised trehalose.

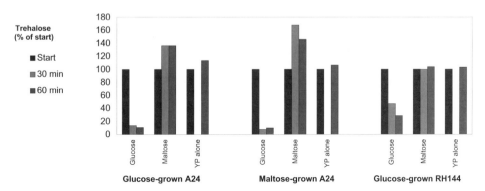

Fig. 3.5 Trehalose mobilisation in yeasts grown on glucose or maltose. A24 lager yeast grown for 4 days at 25°C in YP/2% maltose and containing, respectively, 4.9 mg and 8.7 mg trehalose/g dry mass and the laboratory strain RH144 grown for 4 days at 25°C in YP/2% glucose and containing 91 mg trehalose/g dry mass were added to YP, pH 5, containing 2% glucose, 2% maltose or no added sugar.

3.4 Conclusions

Although maltose, the main sugar in brewers' worts, does not induce a cyclic AMP signal in yeast[5], it caused trehalose mobilisation in all the brewers' strains tested except A143, which did not respond to glucose either. Maltose may trigger trehalase activation and trehalose mobilisation by a signalling system not involving cyclic AMP. Maltose did not mobilise trehalose in commercial baker's yeast (grown on molasses), nor in the maltose-negative RH144 strain nor in the lager yeast, A24, when this was grown to stationary phase on YP/2% glucose or maltose so that its maltose permease activity was relatively small. Future work will test whether trehalose mobilization by maltose requires maltose transport into the cell.

Acknowledgements

We thank Aila Siltala, Eero Mattila and Jari Rautio for technical assistance and the Technology Development Centre, Finland (TEKES) and Oy Panimolaboratorio for financial support.

References

(1)   Quain, D.E. and Tubbs, R.S. (1982) *MBAA Tech. Quart.* **19**, 29–33.
(2)   Callaerts, G., Iserentant, D. and Verchtert, H. (1993) *J. Amer. Soc. Brew. Chem.* **51**, 75–78.
(3)   Thevelein, J.M. and Hohmann, S. (1995) *Trends Biochem. Sci.* **20**, 3–10.
(4)   Van der Plaat (1974) *Biochem. Biophys. Res. Commun.* **56**, 580–587.
(5)   Thevelein, J.M., personal communication.
(6)   Schulze, U., Larsen, M.E. and Villadsen, J. (1995) *Anal. Biochem.* **228**, 143–149.
(7)   Londesborough, J. and Vuorio, O.E. (1993) *Eur. J. Biochem.* **216**, 841–848.
(8)   Londesborough, J. and Varimo, K. (1984) *Biochem. J.* **219**, 511–518.
(9)   Hohmann, S. (1997) In *Yeast Stress Responses*, Hohmann, S. and Mager, W.H. (eds), Springer-Verlag, Heidelberg, pp. 112–119.
(10) Majara, M., O'Connor-Cox, E.S.C. and Axcell, B.C. (1996) *J. Amer. Soc. Brew. Chem.* **54**, 149–154.
(11) Benito, B. and Lagunas, R. (1992) *J. Bacteriol.* **174**, 3065–3069.

4 Relationship Between Cell Deterioration and Plasma Membrane ATPase of *Saccharomyces cerevisiae*

TOSHIHIKO TAKEUCHI and TAKEO IMAI

Abstract   In brewing, yeast vitality is one of most important factors in the production of high quality beer. However, it has been difficult to assess yeast vitality accurately because changes in yeast vitality that appear during beer brewing are very subtle. A sufficiently sensitive method was developed by measuring intracellular pH (plasma membrane ATPase activity) which plays an important role in yeast growth and fermentation. In order to measure proton extrusion from brewer's yeast under physiological conditions, a method determining intracellular pH at low pH was devised. This new method was named intracellular pH or ICP. The intracellular pH of brewer's yeast was measured using a fluorescent reagent and a spectrofluorophotometer. Investigations including comparisons with acidification power test were undertaken. As a result, this method was able to assess subtle differences in the vitality of yeast cells which occur in practical brewing conditions. Since it was now possible to assess cell vitality in brewery yeast, we have also investigated the relationship between yeast vitality and flavour, and found that subtle differences in yeast vitality greatly affected the flavour of beer. Since the plasma membrane ATPase regulates the intracellular pH in yeast the data herein also suggest that plasma membrane ATPase is functionally one of the most important proteins in brewery yeast for the production of high quality beer.

4.1   Introduction

To brew high quality beer it is necessary to determine the activity of the yeast to be used. However, in the brewing process, differences in yeast activities are very subtle, and difficult to differentiate clearly. We have therefore developed a new method for yeast activity assessment using recent advances in yeast physiology and related technologies. It has been demonstrated that intracellular pH and proton pumps play an important role in fermentation regulation[1–3] and cell proliferation activity [4,5]. We have developed a new technique, the intracellular pH method (ICP), which uses intracellular pH as a simple measure of proton extrusion activity, and can be utilised to assess subtle differences in cell vitality. This paper decribes both the ICP method and its application in the brewery.

4.2   Materials and methods

4.2.1   *Yeast strain*

Commercial lager brewing yeasts were used in this study.

4.2.2   *Fermentation performance*

Fermentation was carried out at 8°C in 0.5 l tall-tubes (125 cm × 2.5 cm i.d.). Wort (11°Plato) was air saturated and pitched with yeast at the rate of 3.5g (wet wt)/l. The sugar consumption rate of the yeast was measured after storing the cells at 1°C for various time periods. Sugar consumption was measured during the first 3 days of wort fermentation in a fermenter. Specific gravity was monitored at intervals during fermentation using a digital densitometer (Paar DMA48). Yeast viability was estimated by staining the cells with fluorescein diacetate[6].

4.2.3   *Acidification power test*

The procedure recently described by Kara, Simpson and Hammond[7] was used, which is similar to that described originally by Opekarova and Sigler[8] with the exception that an incubation temperature of 25°C was used in preference to 30°C.

4.2.4   *Measurement of intracellular pH*

The procedure using a spectrofluorometer or microscopic image processor was performed according to the method described by Imai *et al.*[9–11].

4.2.5   *Materials*

5- and 6-Carboxyfluorescein and its acetic ester were obtained from Molecular Probes Inc.; MES (2-(*N*-morpholino)ethanesulfonic acid) and DMSO (dimethyl sulfoxide, LumisonalR) were obtained from Dojindo Laboratories Co., Ltd.

4.3   Results and discussion

4.3.1   *Correlation of the ICP value with fermentation performance*

The intracellular pH (using the ICP method) and the sugar consumption rate were both measured for yeast populations obtained after storage of the cells at 1°C for various time periods[9].

Sugar consumption was measured during the first 3 days of wort fermentation in a 0.5 l laboratory fermenter. The fluctuation in fermentation performance could be clearly predicted by the ICP method results (Fig. 4.1).

4.3.2   *Comparison between the acidification power test and the ICP method*

The acidification power test, which is a very simple assay to perform, has been applied to evaluate yeast vitality. This test was compared with the ICP method using sample yeast populations from practical brewing conditions and the same yeast samples were assessed using both methods. The results show that when the intracellular pH is less than 5.7, the value obtained by the acidification power test is proportional to the intracellular pH obtained by the ICP method (Fig. 4.2). Yeast with an intracellular pH less than 5.7 had a very low proliferation activity[9–11]. However, it should be noted that although the acidification power value is constant when the intracellular pH is greater than 5.7, there are distinct differences in cell proliferation activity from pH 5.7

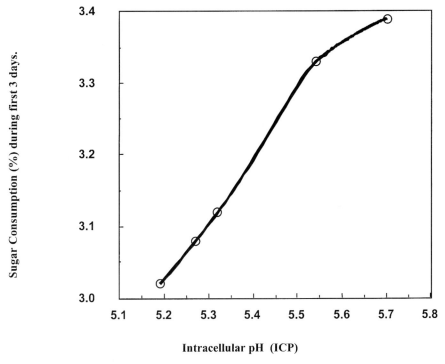

Fig. 4.1   Relationship between intracellular pH and fermentation performance[3].

to 6.3 (Fig. 4.2). Next, the number of yeast samples with different intracellular pHs under practical brewing conditions was investigated (Fig. 4.3). The number of yeast populations possessing intracellular pHs that could be measured by the acidification power test was found to be very low, less than 10% of the total yeast present under practical brewing conditions. Therefore, compared with the acidification power test, the ICP method can be used to measure a much wider range of yeast vitality, such as those actually observed during the brewing process and used to indicate different fermentation performances.

4.3.2.1   *Intracellular pH changes in yeast during storage measured by the ICP method.* The intracellular pH changes measured by the ICP method after storage at various temperatures was investigated. Of course yeast should be stored in a cold room, but 4°C which is a common cold room temperature is here found to be a very dangerous temperature for yeast storage (Fig. 4.4).

4.3.3   *The medium length chain fatty acids and protease activity in beer and beer foam*

After investigating the fermentation performance, we then investigated the relationship between yeast vitality and the quality of the beer produced. Beer foam was measured by the method of Ross and Clark and also was clearly observed to decrease according to yeast vitality (Fig. 4.5). The medium length chain fatty acids were also

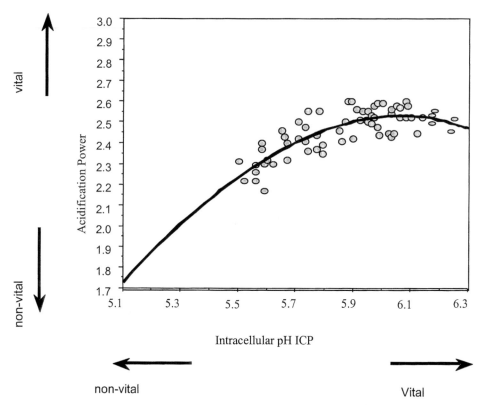

Fig. 4.2 Relationship between the acidification power test and the ICP method[3].

examined, and the sum concentration of middle chain fatty acids was clearly observed to increase with decreasing yeast vitality[12].

In addition, the protease activities in different beers brewed using yeast with different vitality were investigated. It was found that the increase in protease activity depends on the extent of yeast deterioration[12]. It has been known that the effect of wort composition on beer foam values is greater than that of the yeast strain[13] and that yeast handling is very important to improve beer foam quality[14,15]. The data described here support these findings.

4.3.3.1 *The importance of intracellular pH.* Intracellular pH is a crucial factor in yeast physiology for several reasons. Intracellular pH is regulated by plasma membrane ATPase, which is essential for yeast growth and generates the trans-membrane proton gradient. This trans-membrane proton gradient is the driving force for the uptake of nutrients such as maltose and amino acids, which are transported in association with protons. Also, intracellular pH is responsible for upstream regulation of key enzymes in glycolysis and gluconeogenesis. Since these enzymes are regulated by cascade reactions of cAMP dependent protein kinases (e.g. phosphorylase, glycogen synthase, trehalase, fructose-1,6-bisphosphatase and 6-phosphofructo-2-kinase), cAMP plays an important role in these regulations.

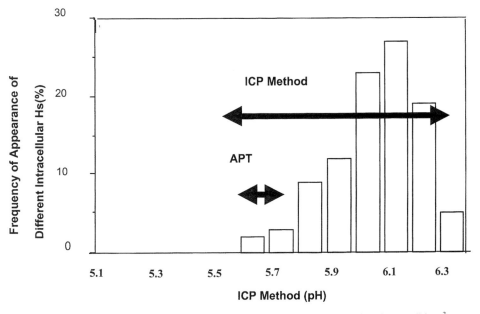

Fig. 4.3   Frequency of appearance of different intracellular pHs during actual brewing conditions[3].

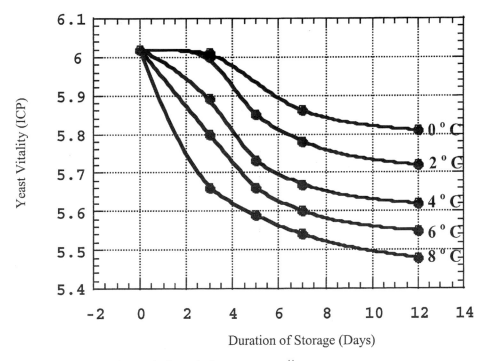

Fig. 4.4   Fluctuation of yeast vitality under low temperatures[11].

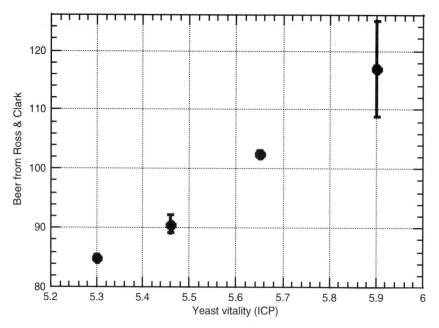

Fig. 4.5   Relationship between beer form and yeast vitality[11].

Moreover it was found that cAMP can be regulated by intracellular pH. From these observations, intracellular pH and proton pump activity (plasma membrane ATPase) are thought to play an important role in yeast growth and glycolysis/ gluconeogenesis. We therefore decided to use intracellular pH and proton pump activity to develop a more highly sensitive method than before. The concept behind the ICP method is to use proton extrusion activity as a measure of yeast vitality. In order to estimate the proton extrusion activity of yeast, measurement of intra- cellular pH was carried out under acidic conditions. It was found that a cell possessing a high proton extrusion activity showed high vitality, and a cell possess- ing a low proton extrusion activity showed low vitality. This demonstrates that proton extrusion activity can be estimated by measuring intracellular pH. To our knowledge, this study is the first to indicate that proton extrusion activity is a key factor in yeast performance under brewing conditions, and that measurement of intracellular pH can be used as a highly sensitive method for assessing proton extrusion, and thus yeast vitality.

4.4   Conclusion

The subtle differences in yeast vitality can be assessed by the ICP method, which is based on plasma membrane ATPase in yeast. The decrease in the ICP value was a deterioration process including PMA. Using this method it is possible to maintain yeast vitality above 6.0 (pH unit) in actual brewing plants.

Acknowledgements

We would like to thank the management of Kirin Brewery Co., Ltd., for permission to publish this work and wish to express our sincere thanks to technical assistants and a number of colleagues for their great help in our work.

References

(1) Noshiro, A., Purwin C., Laux, M., Nicolay, K., Scheffers, W. A. and Holzer, H. (1987) Mechanism of stimulation of endogenous fermentation in yeast by carbonyl cyanide m-chlorophenylhydrazone. *J. Biol. Chem.* **262**, 14154–14157.

(2) Purwin,C., Nicolay, K., Scheffers, W. A. and Holzer, H. (1986) Mechanism of control of adenylate cyclase activity in yeast by fermentable sugars and carbonyl cyanide m-chlorophenylhydrazone. *J. Biol. Chem.* **261**, 8744–8749.

(3) Thevelein, J.M., Beullens, M., Honshoven, F., Hoebeeck, G., Detremerie, K., Den Hollander, J.A. and Jans, A.W.H. (1987) Regulation of the cyclic AMP level in the yeast *Saccharomyces cerevisiae*: intracellular pH and the effect of membrane depolarizing compounds. *J. Gen. Microbiol.* **133**, 2191–2196.

(4) Portillo, F. and Serrano, R. (1989) Growth control strength and active site of yeast plasma membrane ATPase studied by site-directed mutagenesis. *Eur. J. Biochem.* **186**, 501–507.

(5) Serrano, R., Kielland-Brandt, M.C. and Fink, G.R. (1986) Yeast plasma membrane ATPase is essential for growth and has homology with $(Na^+ + K^+)$, K^+- and Ca^{2+}-ATPase. *Nature* **319**, 689–693.

(6) Chilver, M.J., Harrison, J. and Webb, T.J.B. (1977) Use of immunofluorescence and viability stains in quality control. *J. Amer. Soc. Brew. Chem.* **36**, 13–18.

(7) Kara, B.V., Simpson, W.M. and Hammond, R.M. (1988) Prediction of the fermentation performance of brewing yeast with the acidification power test. *J. Inst. Brew.* **94**, 153–158.

(8) Opekarova, M. and Sigler, K. (1982) Acidification power: indicator of metabolic activity and autolytic changes in *Saccharomyces cerevisiae*. *Folia Microbiol.* **27**, 395–403.

(9) Imai, T., Nakajima, I. and Ohno, T. (1994) Development of a new method for evaluation of yeast vitality by measuring intracellular pH. *J. Amer. Soc. Brew. Chem.* **52**, 5–8.

(10) Imai, T. and Ohno, T. (1995) The relationship between viability and intracellular pH in the yeast *Saccharomyces cerevisiae*. *Applied and Environmental Microbiology* **61**, 3604–3608.

(11) Imai, T. and Ohno, T. (1995) Measurement of yeast intracellular pH by image processing and the change it undergoes during growth phase. *J. Biotech.* **38**, 165–172.

(12) Back, W., Imai, T., Forster, C. and Narziss, L. (1998) Hefevitalität und Bierqualität. *Monatsschrift für Brauwissenschaft* **51**, 189–195.

(13) Narziss, L., Miedaner, H. and Gresser, A. (1983) Heferasse und Bierqualität. Der Einfluß der Heferasse auf die Bildung der niederen freien Fettsäuren während der Gärung sowie auf das Niveau der Schaumwerte im Bier. *Brauwelt* **33**, 1354–1357.

(14) Narziss, L., Reicheneder, E. and Voigt, J. (1994) Technologische Faktoren zur Beeinfluessung des Bierschaums. *Brauwelt* **134**, 360–368.

(15) Narziss, L., Reicheneder, E., Barth, D., and Voigt, J. (1994) Technological approach to improve beer foam. *Ferment* **7**, 35–42.

5 Review of Metabolic Activity Tests and Their Ability to Predict Fermentation Performance

DIRK BENDIAK

Abstract Viability and vitality estimates include many indirect tests of metabolic activity that try to predict or correlate how that particular parameter of a yeast sample, taken from the tank of storage yeast prior to pitching, will predict the fermentation performance. A number of specific tests that evaluate a metabolite or cellular activity will be reviewed. These include: (1) monitoring energy levels (ATP levels, adenylate energy change comparisons, ATP/AMP energy pool comparisons, NADH/NADH$_2$ reducing power test, mitochondrial activity); (2) measuring cellular components (CO$_2$ evolution, O$_2$ uptake, ergosterol levels, fatty acid levels, glycogen levels, Mg release rate, yeast protease activity measures); (3) fermentation capacity testing (acidification power test, heat generated, calorimetry, intracellular pH, pH drop, metabolic changes–ethanol, VDK precursors, etc. through the fermentation); (4) cell surface measurements (hydrophobicity, zeta potential); (5) Replication, yeast growth or budding index; (6) flow cytometry; (7) yeast capacitance, and (8) stress indicator measurements (glutathione, iron metabolism, vitamin utilisation, trehalose levels, free radical generation). The ability of the tests to predict fermentation performance is limited. Many are difficult laboratory tests. Simple, preferably on-line, tests may represent the most appropriate assays.

5.1 Introduction

The variation of brewing measurements and control of these parameters is of critical importance in the standardisation and repeatability of the fermentation process to generate consistent beer flavours. Various methods are available for the measurement or estimation of the physiological state of yeast. Vitality and viability estimates have been developed over the years to try to maintain and enhance brewing uniformity and consistency.

These viability and vitality estimates include many indirect tests that try to predict or correlate how a particular parameter of a yeast sample, taken from the tank of storage yeast prior to pitching, will predict the fermentation performance. Viability is defined as a cell's ability to bud and grow, however slowly. Dead is defined as a cell that cannot grow. Vitality, on the other hand, is defined by a continuum of activity of the cell from very active to not active at all (Fig. 5.1).

The vitality tests then should be related directly to yeast growth, yeast physiological changes and fermentation performance. Vitality testing should be simple, rapid and preferably inexpensive so that both large and small breweries can use the test. A new underlying and overriding concern in evaluating many of these tests is the suitability of the method for operator use, enabling control of yeast handling in storage and pitching of a constant amount of healthy yeast into the fermentation.

When testing is performed in the laboratory one has to: collect the sample properly, transport the sample to the laboratory, prepare the sample, run the analysis, record

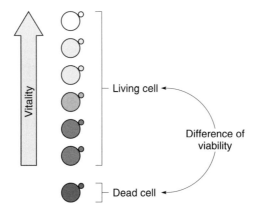

Fig. 5.1 Difference between viability and vitality.

the results, get information to the operator (who is uninvolved and could be unconcerned) and finally act on this information to correct pitching to compensate for suspected vitality differences. An on-line operation can be automated so that the operator, or the system, may respond to data obtained for correcting pitching the yeast to yield more uniform pitch rates.

A number of specific tests that evaluate a metabolite or cellular activity have been reviewed (see Lentini[1] and Imai[2]) and their ability to predict fermentation performance evaluated. These tests have been grouped into seven catagories: energy level tests, cellular component tests, fermentation capacity tests, cell surface tests, replication tests, flow cytometry tests, yeast capacitance tests and stress indicator tests. Each category of test allows the brewer to make the necessary adjustments to the yeast pitching rates and achieve a more controlled fermentation yielding more consistent flavour development (Plate 6).

5.2   Metabolic activity tests

5.2.1   *Monitoring energy levels*

See Table 5.1 for an overview.

5.2.1.1   *Adenosine triphosphate (ATP) levels*[3–7].   The basis of the test is that ATP is present in living cells and absent in dead cells. The method uses bioluminescence of the luciferin and luciferase reaction. When cells are in a similar state of activity then ATP levels are related to viable cell numbers and cell mass. The cell number obtained may be related to many fermentation parameters (growth, and fermentation) and gives a rough estimate of % viability.

5.2.1.2   *Adenylate energy charge (AEC) comparisons*[8,9]. This assessment is based on the ratio of adenylate nucleotides (ATP versus AMP). The nucleotide pool estimates and determines the state of the cells (growth, stress or death phase) and therefore may be used as an indicator of the physiological state of the cell (growth, death, stressed).

Table 5.1   Monitoring energy levels.

Test	Basic principle	Lab. test	General use	Expense
ATP measurement	Measures ATP	Yes	Compares with viable staining	Moderate
Adenyl energy charge	Measures ATP/AMP	Yes	Compares with viable staining	Moderate
Pool measurement	Measures XTP/XMP ratios	Yes	Stress indicator	Very
NADH measurement	Measures NADH	Yes	Compares with viable staining	Moderate
Reducing power	Measures NADH/NAD	Yes	Compares with viable staining	Moderate
Mitochondrial activity	Fluorochrome measurement of enzyme activity	Yes	Mitochondrial activity	Moderate
^{31}P NMR	Measures inorganic phosphate	Yes	Compares with viable staining	Moderate

5.2.1.3   *XTP/XD)/XMP energy pool comparisons.* The A/G/C/T pool comparison resembles the magic spot analysis (ppGpp) in *E.coli*, where pool changes are a good indicator of response to stress.

5.2.1.4   *NADH fluorimetry.* For this assay cells are radiated at 366 nm and the fluorescence detected at 460 nm. NADH is an important source of reducing power in the cell and is a function of the cells' metabolic activity and physiological state. Like ATP, NADH will give a rough estimate of % viability.

5.2.1.5   *NADH/NADH$_2$ reducing power test*[10,11]. The balance of reduced and oxidised forms of NADH/NADH$_2$ relates to the physiological state of the cell and the potential reducing power of the yeast.

5.2.1.6   *Mitochondrial activity*[12,13]. Dye staining for the activity of mitochondria using Rhodamine 123 may represent an important analysis for brewing performance.

5.2.1.7   *^{31}P NMR analysis*[14,15]. Intracellular phosphor is measured using NMR analysis. The cytoplasmic pH value of intact cells and the intracellular state of inorganic phosphate are related to viability measures.

5.2.2   *Measuring cellular components*

See Table 5.2 for an overview.

5.2.2.1   *CO_2 evolution.* Using a Warburg manometer CO_2 release can be measured. One disadvantage of this technique is that sampling of small volumes can give errors,

Table 5.2 Measuring cellular components.

Test	Basic principle	Lab. test	General use	Expense
CO_2 evolution	CO_2 production	Yes	Not indicative of yeast growth	Very expensive
O_2 uptake	O_2 uptake into yeast	Yes	Good for populations of >90% viability Indicates fermentation performance	Very expensive
Glucose uptake	Glucose used	Yes	Not sensitive	Moderate
Ethanol	Ethanol produced	Yes	Not sensitive	Moderate
Ergosterol levels	Major sterol concentration	Yes	Important for understanding oxygen uptake	Moderate
Fatty acid levels	Concentration of fatty acid precursors for sterol production	Yes	Indicator of sterol precursors, sterol production and oxygen requirements	Moderate
Glycogen levels	Glycogen concentration	Yes	Can lack sensitivity	Moderate
Mg release rate	Measure of Mg released	Yes	Related to membrane function	Moderate
Yeast protease activity	Measure of protease released	Yes	Indicates autolysis	Moderate

and measurements are made in a model system, not in a real fermenter. In the fermenter one can monitor progression of fermentation but not react to changes to make suitable corrections where necessary (AE drop, pH drop or yeast growth all occur at the same time).

5.2.2.2 *CO_2 pressure test*[16]. The CO_2 pressure test represents a rapid assessment for monitoring increases in pressure during a 1 h period. However, the instrument required for this assay is expensive and adequate control of reaction temperature and sampling is crucial to obtain reproducible results.

5.2.2.3 *O_2 uptake.* Respiration rates can be assessed by the measurement of oxygen uptake. Fermentation performance is related to oxygen uptake in yeast where the population exhibits viabilities >90%. Where the oxygen uptake rate is low, the O_2 can be increased. Using the BRF yeast vitality test the measurement of oxygen uptake takes 1 h and therefore is relatively rapid. Yeast is pitched at a constant temperature and the pitching rate into aerated wort and oxygen uptake is measured for 1 h. Factors such as yeast acid washing may affect oxygen uptake rate, however, no consistent correlation between this assay and fermentation performance has been observed.

5.2.2.4 *Specific oxygen uptake rate*[17]. Daoud and Searle[17] developed a method which was further modified and is now known as the 'BRF yeast vitality test', which is based on oxygen uptake rates.

5.2.2.5   *Glycolytic flux rates (glucose uptake rate, ethanol production rates)*[18]. Glycolytic flux rates are difficult to monitor, particularly when relatively small changes in glucose or ethanol concentration occur. Although both growing and non-growing cells possess a glycolytic flux rate, sampling small volumes of cell populations can yield errors.

5.2.2.6   *Ergosterol levels*[19]. The measure of internal ergosterol levels gives an indication of sterol production from oxygen. It has been shown that low sterol levels can be compensated for by adequate oxygen provision, making it difficult to predict fermentation performance based on initial sterol concentration. In addition, sterol concentration levels are difficult to monitor.

5.2.2.7   *Fatty acid levels*[20]. The measurement of sterol precursors (fatty acids) has been utilised to predict yeast performance, and may act as an indicator of sterol production and utilisation.

5.2.2.8   *Glycogen levels*[21,22]. Glycogen has been utilised to indicate the viability and adequate storage of cells.

5.2.2.9   *Mg release rate*[23]. The measurement of Mg released from a cell after being placed into wort has been suggested to be a useful indicator of cell vitality. Active cells release Mg while inactive cells release limited levels.

5.2.2.10   *Yeast protease activity measures*[24,25]. The measurement of excreted protease is indicative of death and autolysis of yeast. New more sensitive techniques using specific fluoro probes have been developed to determine this indicator of cell death.

5.2.3   *Fermentation capacity testing*

See Table 5.3 for an overview.

5.2.3.1   *Acidification power test*[26–30]. The acidification power tests measures the reduction in extracellular pH before and after glucose addition. This reduction in pH reflects the extrusion of H+ ions which is concomitant with the excretion of CO_2 and organic acids, and results from K+/H+ exchange and the action of plasma membrane ATPase. The test is an indicator of glycolytic flux rate and involves the Ras-cAMP pathway. Acidification power is a more sensitive indicator than adenylate charge values of vitality.

5.2.3.2   *Cumulative acidification power test*[31,32]. The cumulative acidification power test represents the summation H+ ion extrusion during the acidification power test compensating for the pH log scale.

5.2.3.3   *Titrated acidification test*[33]. For the titrated acidification power test the pH is maintained at 6.3 by addition of NaOH and the amount of base added is reported as an indicator of yeast vitality.

Table 5.3 Fermentation capacity testing.

Test	Basic principle	Lab. test	General use	Expense
Acidification power	Measure of pH before and after glucose addition	Yes	Good for large vitality changes	Moderate
Cumulative acidification power	Summation of change in H+	Yes	Compares with viability	Moderate
Titrated acidification power	Amount of alkaline added to maintain pH of 6.3	Yes	Compares with viability	Moderate
Microcalorimetry	Heat generated from living cell	Yes	No	Very expensive
Intracellular pH (ICP)	Microscopic imaging following exposure to dyes	Yes	Microscopic imaging expensive for routine	Very expensive
pH drop or rise	pH drop of fermentation pH rise of stored yeast slurry	No/Yes	pH monitoring	Cheap
Metabolic changes during fermentation	Fermentation monitoring	Yes	Too late to react	Moderate

5.2.3.4 *Heat generated or micro calorimetry*[34,35]. Microcalorimetry monitors the level of heat production by living cells during growth and fermentation. The test indirectly compares the heat generated with metabolic processes carried out by the cell. The equipment utilised for this assay is expensive but is able to detect minute temperature changes.

5.2.3.5 *Intracellular pH (ICP test)*[36,37]. The intracellular pH or ICP assay utilises fluorescein diacetate FDA and esterised 5,6-carboxyfluorescein (non-fluorescent) to determine intracellular pH changes. The assessment of yeast condition requires microscopic image analysis using spectrofluorophotometery. The technique can detect subtle yeast physiological changes but the imaging equipment is expensive.

5.2.3.6 *pH drop or rise*. pH increase in a yeast slurry in storage indicates autolysis, particularly if the pH is higher than the beer pH from which it came. It is a simple test which can be used by an operator, since pH is a usual check of yeast prior to and during washing. pH measurements can be used to monitor the fermenter, but an unwanted response is usually detected too late to modify pitching and therefore improve fermentation performance.

5.2.3.7 *Monitoring of metabolic changes during fermentation*. Metabolic changes during fermentation include ethanol production, VDK precursor production, protein production and activity and gene expression. The extent, rate and timing of these parameters may reflect yeast cell condition; however monitoring these parameters tends to be retrospective. Response is usually possible only where a multi-brew fermenter is utilised, and more yeast can be added to later brews to correct for under-pitched initial brews.

5.2.4   Cell surface measurements

See Table 5.4 for an overview.

Table 5.4   Cell surface measurements.

Test	Basic principle	Lab. test	General use	Expense
Hydrophibicity	Cell surface composition	Yes	Predicting flocculation	Moderate
Charge	Negative surface compositional groups	Yes	Predicts flocculation and correlates to viability	Moderate

5.2.4.1   *Hydrophobicity measurements*[38].   The measurement of cell surface hydrophobicity may be achieved using a variety of techniques, and has been demonstrated to correlate with flocculation performance and physiological state.

5.2.4.2   *Surface charge zeta potential*[39,40].   Zeta potential is the measurement of yeast cell movement across an electric field and relates to the extent of cell surface negatively charged groups. Zeta potential may relate to methylene blue dye reduction and therefore cell viability.

5.2.5   Replication methods

See Table 5.5 for an overview.

Table 5.5   Replication methods.

Test	Basic principle	Lab. test	General use	Expense
Slide culture	Cell proliferation and non-proliferating cells	Yes	18 h to slow	Cheap
Plating for CFU	Colony forming units (CFU) enumerated	Yes	3 days to slow	Cheap
Short fermentation	Fast fermentation at higher temperature	Yes	24 h to slow	Cheap
Yeast budding Index	% of budded cells measured	Yes	6 to 8 h slow monitoring	Cheap

5.2.5.1   *Slide culture*.   The slide culture is an assessment of cell proliferation based on the capacity of the cells to produce microcolonies on agar over an 18 h period. The microcolonies are enumerated using a microscope. This technique has been recommended by both the American Society of Brewing Chemists and the EBC.

5.2.5.2   *Standard plate count*.   The standard plate count is an assessment of proliferation based on the capacity of cells to produce colonies on agar after 3 days. The colonies are enumerated by eye. Disadvantages include inaccuracies caused by the production of a single colony by one cell, a dividing cell, chains from a chain-forming strain and flocs.

5.2.5.3 *Short fermentation test*[41]. One method for predicting fermentation performance that has been used and that most closely resembles the conditions to which the cells are exposed in fermenter involves the utilisation of a short fermentation of 24 h duration in which various parameters may be examined.

5.2.5.4 *Yeast growth or budding index.* After 6 h in forced fermentation one can measure the % of cells budding and size of the buds they produce. This budding index correlates well with viability. Close fermentation monitoring demonstrates that growth is synchronous for the first two divisions during fermentation.

5.2.6 *Flow cytometry*[43-46]

Fluorescent dyes like Rhodamine 123, oxonol, fluorescein diacetate, carboxy-fluorescein diacetate, propidium iodide, Chemchrome Y, fluorescein-di-β-D-galactopuranoside, C12-DG or other fluorescent probes or antibodies may be used to measure specific activity parameters. This is a rapid and automated but expensive method of cell differentiation.

5.2.7 *Yeast capacitance*[47-49]

Yeast capacity to be charged yields a signal which can be related to cell viability utilising an automated instrument which is known as an Aber Unit. This technique may be used in practice on line to monitor and control for pitching and yeast collection.

5.2.8 *Other stress indicator measurements*

See Table 5.6 for an overview.

5.2.8.1 *Glutathione monitoring*[50]. The measurement of cellular reduced glutathione has been related to viability and may be used as a stress indicator.

Table 5.6 Stress indicator measurements.

Test	Basic principle	Lab. test	General use	Expense
Glutathione	Reduced glutathione	Yes	Stress indicator	Moderate
Iron metabolism	Iron utilisation	Yes	Fermentation oxygen requirement predictor	Moderate
Vitamin utilisation	Vitamin levels used	Yes	Indicator of physiological changes	Moderate
Trehalose levels	Trehalose measurement	Yes	Indicates stress of yeast	Moderate
Free radical generation	Measurement of free radicals	Yes	Indicator of stress	Very expensive
NMR analysis	NMR of active protons	Yes	Measurement of change in yeast condition	Very expensive

5.2.8.2 *Iron metabolism.* Changes in iron levels can indicate possible oxygen transport molecule levels and in turn relate to fermentation performance.

5.2.8.3 *Vitamin utilisation.* Vitamin utilisation indicates physiological changes in the yeasts.

5.2.8.4 *Trehalose levels*[51–53]. The measurement of trehalose levels during yeast storage indicates the stress levels to which the cells have been exposed. Generally trehalose levels tend to be consistent unless the storage regime has been abused or the yeast has been mishandled.

5.2.8.5 *Free radical generation.* Free radical generation indicates the metabolic state of the cell and may also reflect the extent of stress response exhibited by the yeast.

5.2.8.6 *NMR analysis.* NMR analysis may be utilised to differentiate between stressed and non-stressed yeast.

All the techniques summarised have been used in an attempt to predict the capacity of the yeast to perform when pitched into a fermenter, but during the fermentation and storage a large number of other parameters must be considered including

- Pitching rate (since underpitching can lead to hung fermentations)
- Multiple yeast in one brewery, particularly those which are infrequently used
- The uniformity of distribution or dispersion of yeast in the fermenter (pitched throughout wort cooling or for 10 min in the 60 min cycle)
- That yeast growth occurs as indicated by the rate and extent or % of budding cells, cell division and extent and duration of the lag phase
- Individual cell age within a culture will comprise

50%	newly budded first generation cells	one birth scar
25%	generation F1	one birth scar, one bud scars
12.5%	generation F2	one birth scar, two bud scars
6.25%	generation F3	one birth scar, three bud scars
3.125%	generation F4	one birth scar, four bud scars
1.5625%	generation F5	one birth scar, five bud scars

or 98.4375% of cells 5 generations old (or two brewing fermentations) with F6 plus accounting for the remaining 1.65% of cells with six or more bud scars.

- Consistent wort oxygenation, oxygen uptake for all yeast cells within a given population
- That the fermentation temperature was properly controlled to yield the fermentation profile desired
- That the cooling temperature was correct $\pm$ 0.5°C
- That the rate of heating in the fermenter to fermentation temperature was consistent (cellar temperatures constant, summer versus winter cellar temperatures can be different)
- That the nutrients were correct and were typical of the standard wort production process

- That internal pools of energy, and reducing power in the yeast were more than adequate
- That yeast were not suffering from stress for yeast storage and handling procedures
- That yeast were stored at 0 to 2°C for the shortest possible time
- That yeast responded properly to CO_2 inhibition
- Fermenter design
- Yeast response to osmotic pressures
- Flocculation responses.

5.3 Conclusions

In order to control fermentation it is necessary to limit variations in: oxygen levels, viable yeast cell pitching rate, homogeneity in the pitching solution, temperature of cooling, temperature of fermentation and wort nutrient production; generally this yields fermentations that perform in terms of minimal lag, rapidity, repeatability and completion, with consistent pH drop and ester profiles and flavours. Usually the problem brews are a failure in one of the above.

We are asking one test to tell us: all the life history, all the accompanying baggage, and all the minute changes of the yeast physiology as a prediction of fermentation performance. We want this one parameter to tell us if our yeast is OK to reuse and whether it will perform properly in the next brew. I think not.

We still need to understand the relationship of the vitality continuum, vitality measures, and yeast cell population versus fermentation performance as indicated in Table 5.7.

Table 5.7 Dependence of fermentation on vitality.

Vitality continuum	Viability measurement		Fermentation performance
100%	100%		Normal if viable cells pitched
90%	90% OK	10% poor	Close to normal if pitched
80%	80% OK	20% poor	Close to normal if pitched
60%	60% OK	40% poor	Slow inadequate viable cells
40%	40% OK	60% poor	Slow inadequate viable cells
20%	20% OK	80% poor	Inadequate viable cells
0%	0% Dead		No fermentation

Currently the most favoured tests are those which are simple, easy and reasonably cheap for an operation independent of size and include: yeast pH (ICP), budding index, protease, capacitance measurement and, of course, the fluorescent and reductive stains.

Acknowledgements

I wish to thank Molson Canada for the opportunity to present this information.

References

(1) Lentini, A.A. (1993) Review of the various methods available for monitoring the physiological status of yeast: yeast viability and vitality. *Ferment* **6**, 321.

(2) Imai, T. (1999) The assessment of yeast vitality – the past and the future. *Brewers Guardian* **128**(6), 20.

(3) Davis, W.M. and White, D.C. (1980) *Applied Environmental Microbiology* **40**, 539.

(4) Little, K.J. and La Rocco, K.A. (1985) *Journal of Food Protection* **48**, 1002.

(5) Hysert, D.W., Kovecses, F. and Morrison, N.M. (1976) *J. Amer. Soc. Brewing Chem.* **34**, 145

(6) Miller, L.F., Mabees, M.S., Gress, H.S. and Jangaard, N.O. (1978) *J. Amer. Soc. Brewing Chem.* **36**, 59.

(7) Hysert, D.W. and Morrison, N.M. (1977) *J. Amer. Soc. Brewing Chem.* **35**, 253.

(8) Karl, D.H. and Holm-Hansen, O. (1978) *Marine Biology* **48**, 185.

(9) Schimz, K.L. (1980) *Advances in Biotechnology* **1**, 457.

(10) Eisele, A. and Maric, V. (1980) *Advances in Biotechnology* **1**, 335.

(11) Muller, W., Wehnert, G. and Scheper, T. (1988) *Analytica Chemica Acta* **213**, 47.

(12) Shen, H.-Y., Verachtert, H. and Iserentant, D. (1997) *Proc. Brew. Yeast Ferm. Perf. Cong.* **1**, 19.

(13) Lodolo, E.J., O'Connor-Cox, E. and Axcell, B. (1994) *Proc. Inst. Brew* (Cent. S.A. Sect.) **4**, 167.

(14) Furukubo, S., Matsumoto, Y., Yomo, H., Fukui, N., Ashakari, T. and Kakimi, Y. (1997) *Proc. Eur. Brew. Conv. Mastricht*, 423.

(15) Dinsdale, M., Lloyd, D., McIntyre, P. and Jarvis, B. (1999) *Yeast* **15**, 285.

(16) Muck, E. and Narziss, L. (1988) *Brauwelt International*, 61.

(17) Daoud, L.L. and Searle, B.B. (1986) European Brewing Convention Monograph XII: Symposium on Brewing Yeast, p. 108.

(18) Jones, R.P. (1987) *Process Biochemistry*, 118.

(19) Prasad, R. (1985) *Advances in Lipid Research* **21**, 187.

(20) Beaven, M.J., Charpentier, C. and Rose, A.H. (1982) *Journal of General Microbiology* **128**, 1447.

(21) Quain, D.E. and Tubb, R.S. (1983) *Journal of Institute of Brewing* **89**, 38.

(22) Quain, D.E. (1981) *Journal of Institute of Brewing* **87**, 289.

(23) Mochaba, E., O'Connor-Cox, E. and Axcell, B. (1998) *J. Amer. Soc. Brewing Chem.* **56**, 1.

(24) O'Connor-Cox, E., Mochaba, F., Lodolo, E., Majara, M. and Axcell, B. (1997) *MBAA Tech. Quart.* **34**, 306.

(25) Kondo, H., Yomo, H., Furukubo, S., Kawaskai, Y. and Nakatani, K. (1998) *Proc. Conv. Inst. Brew. (Asia Pacific Sect.) Perth* (25), 119.

(26) Kara, B., Simpson, W. and Hammond, J. (1988) *J. Inst. Brew.* **94**, 153.

(27) Willetts, J., Seward, R., Dinsdale, M., Suller, M., Hill, B. and Lloyd, D. (1997) *J. Inst. Brew.* **103**, 79–84.

(28) Mathieu, C., van den Berg, L. and Iserentant, D. (1991) *Proc. Congress EBC* **23**, 273.

(29) Thevelein, J. (1992) *Antonie van Leeuwenhoek* **62**, 109.

(30) Thevelein, J. (1994) *Yeast* **10**, 1753.

(31) Patino, H., Edelen, C. and Miller, J. (1993) *J. Amer. Soc. Brewing Chem.* **51**, 128.

(32) Patino, H., Edelen, C. and Miller, J. (1993) *MBAA Tech. Quart.* **30**, 98.

(33) Iserentant, D., Geenens, W. and Verachtert, H.J. (1996) *J. Amer. Soc. Brewing Chem.* **54**, 110.

(34) O'Toole, D. (1983) *Journal of Applied Bacteriology* **55**, 187.

(35) Perry, B. (1983) *Journal of Applied Bacteriology* **54**, 183.

(36) Imai, T. (1999) *Brewers Guardian* **128**, 20.

(37) Imai, T. (1996) *Proc. Conv. Inst. Brew* **24**, 60.

(38) George, E., Hodgson, J.A. and Smart, K.A. (1995) *Proc. Eur. Brew. Conv.*, 329–336.

(39) Brown, P. (1996) *Proc. 6th Int. Brew. Conf.*, Harrogate, 339–407.

(40) Brown, P. (1997) *Brewer's Guardian* **126**, 39.

(41) Manson, D.H. and Slaughter, J.C. (1986) Proceeding of the 2nd Aviemore Conference in Malting, Brewing and Distillation, Aviemore, p. 297.

(42) Lloyd, D., Moran, C., Suller, M., Dinsdale, M. and Hayes, A. (1996) *J. Inst. Brew.* **102**, 251.

(43) Iserantant, D. *Proc. Int. Symp. Malting and Brewing Technology*.

(44) Donhauser, S., Eger, C., Hubl, T., Schmidt, U. and Winnewisser, W. (1993) *Brauwelt International* **3**, 221.

(45) Hutter, K.-J. (1996) *Brauwelt International* **14**, 52.

(46) Edwards, C., Porter, J. and West, M. (1996) *Ferment* **9**, 288.

(47) Pateman, B. (1997) *Brew. Distill. Int.* **28**, 57.

(48) Carvell, J. (1997) *Ferment* **10**, 261.

(49)   Karvell, J. (1999) *Brew. Distill. Int.* **30**, 17.
(50)   Kolb, N. and Maric, V. (1994) Proc. 6th Eur. Congress of Biotechnology, *Prog. Biotechnol* **9**, 101.
(51)   Guldfeldt, L. and Arneborg, N. (1998) *J. Inst. Brew.* **104**(1), 37.
(52)   Van Landschoot, A. (1998) *Belg. J. Brew. Biotechnology* **23**, 23.
(53)   Peter, R. (1998) *Proc. Inst. Brew. Asia Pacific Section* **25**, 200.

6 Predicting Fermentation Performance Using Proton Efflux

RUKHSANA SIDDIQUE and KATHERINE SMART

Abstract   Viability and vitality are the two terms used to describe pitching yeast quality. Yeast viability is a measure of the number of living cells, while yeast vitality indicates the level of yeast activity and represents the physiological state of the viable cell population during assessment. Many methods are available for assessing yeast viability and vitality; however, most of these methods either lack the sensitivity required to assess populations following extreme physiological stress or monitor only one aspect of metabolism which may not correlate with subsequent fermentation potential. Assessing yeast vitality by proton efflux using the acidification power test yields an acceptable correlation with fermentation performance and is relatively rapid and simple to use. The acidification power test is based on the ability of a yeast population to acidify the surrounding medium before (spontaneous) and after (substrate induced) the addition of a metabolisable sugar such as glucose. The change in pH observed represents the 'total acidification power' of the yeast culture. Since the test determines the extent of proton efflux from the cell, it reflects the plasma membrane trans-membrane potential. Despite the potential advantages of this method, there is a requirement for development to enhance both the sensitivity and reproducibility of the assay. In this chapter we suggest an improved method for determining the extent of proton efflux from yeast cultures.

The proton efflux of brewing yeast strains (lager, ale and cider) in different physiological states has been examined and compared with more conventional means of both viability and vitality assessment using brightfield and fluorescent dye reduction and plate count techniques.

The improved acidification power test correlated well with fermentation performance and the dye reduction assay. In particular, it is suggested that the close correlation between the improved proton efflux assay and the alkaline methylene violet procedure demonstrates potential for the prediction of yeast fermentation performance for this new method. It is suggested that the improved acidification power assay represents a useful alternative to methylene blue for evaluating yeast viability for brewing and cider yeast.

6.1 Introduction

The acidification power test was first devised by Opekarova and Sigler[1]. The extracellular pH value of a resting suspension of yeast in distilled water ranges from 4 to 8 depending on the yeast strain and the previous history of that strain[2,3]. On addition of metabolisable carbohydrate (glucose, maltose) to the resting yeast suspension, the pH drops[4]. Both phenomena are an expression of the tendency of the yeast to establish a fixed ratio between extracellular and intracellular hydrogen ion concentrations[2,3]. The energy to maintain this pH difference is, in the absence of added fermentable carbohydrate, provided by the cells' endogenous energy reserves (glycogen and/or trehalose). In the presence of a fermentable carbohydrate such as glucose, the energy is provided by both endogenous and exogenous energy sources, thus reflecting glycolytic activity[1–3]. Brewer's yeast must synthesise an adequate quantity of sterol in order to grow, and internal energy reserves (glycogen) fuel this synthesis[2]. Fermentation of brewery wort is the result of the metabolism of wort sugars via the Embden–Meyerhof–Parnas (or glycolytic) pathway. Therefore, it should be possible, to relate the potential of a given pitching yeast for growth to the spontaneous acidification power (ΔAP10) and the fermentative activity of the yeast with the glucose-induced acidification power (ΔAP20). The deduction of these two values from the initial pH gives a value for the acidification power (ΔTAP) of the yeast[5]. Here the development of the acidification power test is reported.

6.2   Aims of the investigation

The aims are to determine the relationship between physiological state and acidification power and to develop adequate controls for the acidification power test.

6.3   Materials and methods

6.3.1   *Yeast strains and growth conditions*

Lager (KS1), ale (2593) and cider (TC16) brewing yeast strains were utilised in this study. The lager strain KS1 (formerly BB5) was obtained from Bass Brewers Ltd., Burton-on Trent, UK; the ale strain, 2593, was obtained from the National Collection of Yeast Cultures, Norwich, UK; and the cider strain, TC16, was obtained from Taunton Cider PLC, Taunton, UK. Strains were stored on beads at $-80°$C in YPD containing 20% (w/v) glycerol. Stock cultures were maintained on YPD consisting of 1% (w/v) yeast extract, 0.5 neutralised bacteriological peptone, and 1% glucose solidified with 1.5% agar (w/v). All media were autoclaved immediately after preparation at 121°C and 15 psi for 15 min.

Yeasts were grown aerobically to the required cell density at 25°C by shaking in an Erlenmeyer flask (250 ml) at 120 min^{-1}. Cell growth was monitored and measured by optical density at 600 nm wavelength using a spectrophotometer.

6.3.2   *Yeast cell population status*

To achieve stationary and exponential physiological states, yeast cells were grown aerobically in YPD at 25°C. Cells were then harvested by centrifugation (at 3000 × g for 5 min, at 4°C) washed three times in cold, deionised sterile water, and diluted to give a final concentration of 1×10^7 cells/ml. The cell suspension was kept on ice until required. To achieve non-viable cell populations, yeast cells were heat treated at 65°C for 2 h. Stressed populations were obtained by starving yeast cells in water for 24 h, 48 h and 72 h, stored statically at 4°C.

6.3.3   *Acidification power test*

The GAP, or the glucose acidification power test, was undertaken according to a modified version of the method of Kara *et al.*[4]. In a universal bottle sterile deionised water (15 ml) was continuously stirred at 150 rpm at 25°C for 5 min to allow for pH probe (Camlab Limited, Nuffield Road, Cambridge CB4 1TH, UK) equilibration. The spontaneous proton efflux was monitored for 10 min (ΔAP10) prior to the addition of glucose solution (5 ml of 20.2%, w/v). The proton efflux was then monitored for a further 10 min (ΔGAP).

The MAP, or the maltose acidification power test, and WAP, the water acidification power test, were performed according to the same method as described for the glucose acidification power test; however, glucose was replaced by maltose solutions (5 ml of 20.2 w/v) and water (5 ml sterile, deionised).

For the AAP, or sodium azide acidification power test, after recording the pH for 10 min (ΔAP10), glucose solution (5 ml of 20.2 w/v) was immediately added to the yeast suspension and the pH recorded for 5 min. Then, 5 ml of 1M sodium azide solution was immediately added and the pH was recorded for a further 5 min (ΔAAP).

The BAP, or base acidification power, was based on heat-treated cells. Harvested and washed yeast cell (1 $\times$ 10^7 cells/ml) were killed by heat treatment at 65°C for 2 h. After recording the pH for 10 min (ΔAP10), glucose solution (5 ml of 20.2 w/v), maltose solution (5 ml of 20.2 w/v) or water (5 ml sterile, deionised) was added and the pH was then recorded for a further 10 min (ΔBAP).

For each sample of yeast suspension examined, five replicate acidification powers were recorded. The total acidification power (ΔTAP) was calculated by subtracting the final pH value (after 20 min) from the initial pH of the water (ΔAP0).

6.4   Results and discussion

6.4.1   *Does acidification power indicate yeast viability?*

When yeast is repeatedly used for a series of primary fermentations its efficiency will gradually deteriorate[6,7]. Therefore, it is common practice for brewers to measure yeast viability and condition prior to pitching, as this allows adjustments to be made to the quantity of yeast pitched. In order to maintain this efficient process a rapid and accurate determination of yeast viability in the brewery is an absolute necessity[8]. The method utilised should be able to distinguish between live and dead yeast cells efficiently and without ambiguity. Over the years, several methods have been developed to measure yeast viability based on the yeast cell's ability to grow and replicate, the permeability of the cell membrane and the metabolic activity of the cell. Methylene blue dye retention has been internationally recommended as a standard procedure for measuring yeast viability by the European Brewing Convention[9] and the American Society of Brewing Chemists[10]. But, the non-reliability of the methylene blue dye retention assay has been highlighted by many authors[11–14]. Furthermore it has been postulated that this may be due to the presence of non-reducible impurities in the dye which lead to erroneous results[14]. In addition, for dead cell populations methylene blue may indicate viabilities as high as 30 to 40%[11,13,15,16].

There is therefore a requirement for the development of a method for yeast viability assessment. It has been suggested that the acidification power test may represent a useful alternative[14]. To investigate this, three brewing yeast strains were utilised and populations of viable (exponential and stationary) and non-viable (heat killed) cells were obtained. It was observed that the acidification power test did distinguish between live and dead cells (Table 6.1). However, unlike other methods, non-viable cell populations did not have a value of zero. The acidification powers for healthy exponential, live cell populations were strain dependent, with no absolute maximum value corresponding to 100% viability (Table 6.1). The viability of heat treated cell populations using the acidification power (Table 6.1) resulted in some proton efflux, largely during $\Delta AP10$. However, when the glucose was added no increase in proton efflux occurred.

Table 6.1 Acidification power of (a) viable exponential and (b) non-viable (2 h heat killed) populations.

Strain	Lager (KS1)	Ale (2593)	Cider (TC16)
(a) Viable			
AP10	0.89 ± 0.24	0.29 ± 0.24	1.04 ± 0.10
GAP	1.53 ± 0.09	1.27 ± 0.29	1.35 ± 0.14
TAP	2.42 ± 0.23	1.72 ± 0.10	2.40 ± 0.04
(b) Non-viable			
AP10	0.72 ± 0.28	0.67 ± 0.18	0.42 ± 0.10
GAP	0.13 ± 0.05	0.08 ± 0.20	0.02 ± 0.14
TAP	0.59 ± 0.28	0.75 ± 0.10	0.44 ± 0.04

6.4.2   Does acidification power indicate yeast vitality?

It has been suggested by Mathieu et al.[17] that the acidification power test (APT) allows an estimation of the physiological condition of yeast. The capacity of acidification power to distinguish between cell populations exhibiting different physiological states was examined, and it was observed that this assay distinguishes between vital exponential, vital stationary phase and non-vital stored cell populations (Tables 6.2–6.4). It has been suggested that $\Delta AP10$ values reflect the intracellular carbohydrate reserves[6]. $\Delta AP10$ was observed to be strain and growth phase dependent (Table 6.2). AP10 did not reflect physiological state per se. This is not surprising since the relationship between intracellular carbohydrate reserves and physiological stress is likely to be complex (Table 6.2). However, it is suggested that the proton efflux which occurs in water should be termed 'resting' efflux. AP20, which reflects the

Table 6.2 Capacity of spontaneous proton efflux ($\Delta AP10$) which reflects the physiological state.

Strain	Lager (KS1)	Ale (2593)	Cider (TC16)
Exponential phase	0.89 ± 0.24	0.29 ± 0.24	1.04 ± 0.10
Stationary phase	0.55 ± 0.15	0.15 ± 0.23	0.11 ± 0.17
Stored 24 hrs	0.51 ± 0.27	0.40 ± 0.16	0.17 ± 0.07
Stored 48 hrs	0.16 ± 0.09	1.92 ± 0.22	0.35 ± 0.26
Stored 72 hrs	0.22 ± 0.08	2.39 ± 0.06	0.42 ± 0.16
Killed	0.72 ± 0.28	0.67 ± 0.18	0.42 ± 0.10

TAble 6.3 Capacity of glucose induced proton efflux (ΔGAP) to reflect physiological state.

Strain	Lager (KS1)	Ale (2593)	Cider (TC16)
Exponential phase	1.53 ± 0.09	1.27 ± 0.29	1.35 ± 0.14
Stationary phase	1.24 ± 0.05	1.42 ± 0.07	1.53 ± 0.13
Stored 24 hrs	0.67 ± 0.35	1.41 ± 0.40	0.97 ± 0.14
Stored 48 hrs	0.36 ± 0.24	0.82 ± 0.20	0.98 ± 0.09
Stored 73 hrs	0.65 ± 0.07	1.15 ± 0.06	1.17 ± 0.16
Killed	0.13 ± 0.28	0.75 ± 0.10	0.44 ± 0.04

glycolytic activity of the cell, is also strain and growth phase dependent (Table 6.3 and Fig. 6.1). Glycolytic activity as represented by ΔGAP reflected the physiological state for the form of stress imposed on the strains in this study. These findings corroborate the observations of Smart and coworkers[14].

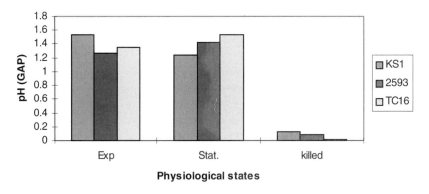

Fig. 6.1   Substrate induced proton efflux (ΔGAP) differentiates between live and dead cell populations.

In terms of vitality measurements, it was observed that the responses were strain dependent. For the lager strain, KS1, the TAP for stressed populations was significantly less than that observed for non-stressed populations. However, the stressed populations appeared to exhibit similar total acidification powers to those exhibited by heat treated non-viable populations. The reasons for this are not known but indicate that this means of assessment may be unsuitable for this strain. In terms of the intracellular carbohydrate reserves (ΔAP10) for this strain, again this value did not relate to the physiological state of the cell. However, the ΔAP20 or glycolytic activity differed for non-viable, stressed and non-stressed populations, indicating that this parameter may represent a potential vitality test for this strain, corroborating the findings of Smart *et al.*[14].

For the ale strain (2593) and the cider strain (TC16), the ΔTAP adequately distinguished between cell populations which were viable, non-stressed and stressed (starved aerobically at 25°C) but not stored (starved statically at 4°C) populations. This result appeared to be related to the ΔAP10 or reserve carbohydrates which apparently increased during storage. This is surprising since it has been suggested that prolonged storage results in the depletion of glycogen and other reserve carbohydrates[6,18]. The ΔAP20, which represents the glycolytic activity of the cell, appears to distinguish between populations of cider (TC16) and lager (KS1) yeast on the basis of physiological condition, but not those of the ale strain (2593).

It is possible that one of the primary reasons for the variability observed is the lack of adequate controls for this assessment technique, particularly given the high degree of variability observed for the killed cell populations of the same strain (Tables 6.4 and 6.5).

Table 6.4 Capacity of total acidification power (ΔTAP) to reflect physiological state.

Strain	Lager (KS1)	Ale (2593)	Cider (TC16)
Exponential phase	2.42 ± 0.23	1.72 ± 0.10	2.40 ± 0.04
Stationary phase	1.80 ± 0.12	1.57 ± 0.16	1.41 ± 0.24
Stored 24 hrs	1.18 ± 0.54	1.81 ± 0.26	0.92 ± 0.22
Stored 48 hrs	0.45 ± 0.30	2.75 ± 0.37	1.33 ± 0.27
Stored 72 hrs	0.69 ± 0.40	3.54 ± 0.11	1.61 ± 0.25
Killed	0.59 ± 0.28	0.75 ± 0.1	0.44 ± 0.04

Table 6.5 Proton efflux generated by (a) exponential and (b) stored (4°C, 72 h) yeast populations following exposure to glucose (ΔGAP), water (ΔWAP), sodium azide (ΔAAP), heat treatment (ΔBAP) and maltose (ΔMAP).

Strain	Lager (KS1)	Ale (2593)	Cider (TC16)
(a) Exponential population			
GAP	1.53 ± 0.09	1.27 ± 0.29	1.35 ± 0.14
WAP	0.12 ± 0.09	0.08 ± 0.10	0.07 ± 0.01
AAP	0.26 ± 0.17	0.32 ± 0.17	0.20 ± 0.13
BAP	0.13 ± 0.28	0.75 ± 0.1	0.44 ± 0.04
MAP	0.19 ± 0.01	0.12 ± 0.10	0.19 ± 0.25
(b) Stored population			
GAP	0.65 ± 0.07	1.15 ± 0.06	1.17 ± 0.16
WAP	0.08 ± 0.07	0.10 ± 0.08	0.03 ± 0.02
AAP	0.62 ± 0.06	0.44 ± 0.06	0.45 ± 0.08
BAP	0.04 ± 0.04	0.05 ± 0.04	0.10 ± 0.06
MAP	0.15 ± 0.00	0.42 ± 0.03	0.07 ± 0.07

6.4.3 What controls should the acidification power test include?

The acidification power test does distinguish between live and dead cells, but this assay requires calibration for each strain utilised. It is suggested that the acidification power should be calibrated to citrate methylene violet[14] to establish the values representing 100% viability. Although the ΔGAP can distinguish between live and dead cells, it is clear that background proton efflux or pH fluctuations occur within the test, so that for a dead cell population proton efflux is apparent. It is suggested that a control be introduced which eliminates this background efflux.

This could be achieved by determining the ΔGAP for heat killed (BAP) and sodium azide treated (AAP) cell populations (Table 6.5). Sodium azide treatment resulted in a decreased proton efflux which was strain dependent. However, the efflux was also dependent on the physiological state of the populations prior to treatment, indicating that this did not represent an adequate negative control. BAP constitutes the 'passive' efflux of protons from non-viable cells and thus represents zero viability. It is suggested that ΔGAP $-$ ΔBAP should be calculated to eliminate passive efflux and pH probe fluctuations (Table 6.6). The ΔGAP proton efflux also comprises both active (glucose induced) and resting efflux as a result of immersion in a solution. It is suggested that the proton efflux which occurs in water be monitored for 0 to 10 min (ΔAP10) and 10 to 20 min (WAP) (Table 6.5). By subtracting ΔWAP from ΔGAP, the 'active' proton efflux is calculated (Table 6.6 and Fig. 6.2). By subtracting ΔWAP and ΔBAP from ΔGAP the 'glucose induced proton efflux' is calculated (Table 6.6 and Fig. 6.3).

The main carbohydrate in wort is not glucose but maltose[19]. It has been suggested that the proton efflux generated by maltose utilisation (MAP) may be a more appropriate measure of brewing yeast vitality (Table 6.5). However, the duration of incubation was not sufficient to allow detectable levels of proton efflux. This method

Table 6.6   Application of controls (ΔBAP and ΔWAP) to the active proton efflux generated by exposure to glucose (ΔGAP) for (a) exponential phase and (b) stored (4°C, 72 h) yeast populations.

Strain	Lager (KS1)	Ale (2593)	Cider (TC16)
(a) Exponential population			
GAP	0.89 ± 0.24	0.29 ± 0.24	1.04 ± 0.10
GAP-ΔWAP	1.40 ± 0.18	1.22 ± 0.14	1.28 ± 0.15
GAP-ΔBAP	1.49 ± 0.13	1.26 ± 0.22	1.30 ± 0.17
(b) Stored population			
GAP	0.65 ± 0.07	1.15 ± 0.06	1.17 ± 0.16
GAP-ΔWAP	0.56 ± 0.03	1.05 ± 0.02	1.14 ± 0.18
GAP-ΔBAP	0.61 ± 0.11	1.10 ± 0.05	1.07 ± 0.18

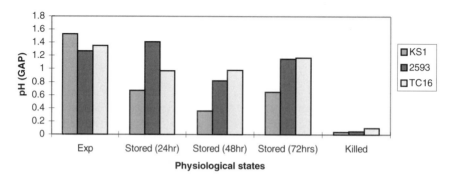

Fig. 6.2 Substrate induced proton efflux (ΔGAP) differentiates between vital and non-vital cell populations.

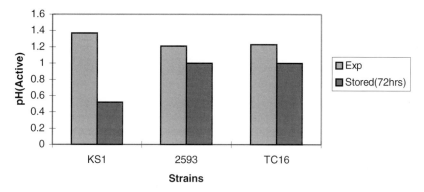

Fig. 6.3 Glucose induced proton efflux.

may be more appropriate for monitoring the cell physiological state during propagation, where cells are grown on and utilise maltose.

6.5   Conclusions

The viability and vitality of brewing yeast cultures affect fermentation performance and subsequently beer quality[14]. Since the number of viable cells is important to the brewer in terms of accurately calculating pitching rates, and the vitality of the living cells is important in predicting yeast activity in the fermentation, it is necessary that accurate, reproducible yet simple and inexpensive assays be developed to predict the physiological state of the cell.

It is suggested that the proton efflux resulting from glycolytic activity (ΔAP20) which is one of many measurements made during the acidification power test to predict cell vitality be utilised. AP20, however, requires method development, since the results obtained are strain specific, and controls which did not exist for the previously published method[4] have yielded inconsistent results. This remains the subject of future investigations.

Acknowledgement

The authors would like to gratefully acknowledge Dr. Gilbert Vandenput for funding the M.Phil. study of R.S. at Oxford Brookes University.

References

(1)   Opekarova, M. and Sigler, K. (1982) Acidification power: indicator of metabolic activity and autolytic changes in *Saccharomyces cerevisiae*. *Folia Microbiology* **27**, 395–403.
(2)   Sigler, K., Kotyk, A., Knotkova, A. and Opekarova, M. (1981) Factors governing substrate-induced generation and extrusion of protons in the yeast *Saccharomyces cerevisiae*. *Biochemica et Biophysica Acta* **643**, 572–582.

(3) Sigler, K., Kotyk, A., Knotkova, A. and Opekarova, M. (1981) Processes involved in the creation of buffering capacity and in substrate-induced proton extrusion in the yeast *Saccharomyces cerevisiae*. *Biochemica et Biophysica Acta* **643**, 583–592.

(4) Kara, B.V., Simpson, W.J. and Hammond, J.R.M. (1988) Prediction of the fermentation performance of brewing yeast with the acidification power test. *J. Inst. Brew.* **94**, 153–158.

(5) Mansure, J.J.G., Panek, A.D., Crowe, L.M. and Crowe, J.H. (1994) Trehalose inhibits ethanol effects on intact yeast cells and liposomes. *Biochimica et Biophysica Acta* **1191**, 309–316.

(6) Boulton, C. (1991) Yeast management and the control of brewing fermentation. *Brewers Guardian* **120**, 25–29.

(7) Smart, K.A. (1996) Nutritional requirements and performance of yeast. *Eur. Brew. Conv. Monogr. XXIV symp. Immobilized Yeast Appl. Brew. Indus.* 146–157.

(8) Lentini, A. (1993) A review of the various methods available for monitoring the physiological status of the yeast and vitality. *Ferment* **6**, 321–327.

(9) European Brewing Convention (1977) Analytica Microbiologica: Method 2.2.2.3.5. *J. Inst. Brew.* **83**, 109–118.

(10) ASBC (1976) *Methods of analysis*, 7th edn, *Yeast*, Part 3A, Dead Cell Stain, American Society of Brewing Chemists, St Paul, MN.

(11) King, L.M., Schisler, D.O. and Ruocco, J.J. (1980) Epifluorescent method for detection of nonviable yeast. *J. Amer. Soc. Brew. Chem.* **39**, 52–54.

(12) McCaig, R. (1990) Evaluation of the fluorescent dye 1-anilino-8-naphthalene sulphonic acid for yeast viability determination. *J. Amer. Soc. Brew. Chem.* **48**, 22–25.

(13) Pierce, J. (1970) Institute of Brewing Analytical Committee: Measurement of Yeast Viability. *Journal of the Institute of Brewing* **76**, 442–443.

(14) Smart, K.A., Chambers, K.M., Lambert, I., Jenkins, C. and Smart, K.A. (1999) Methylene violet staining for yeast viability and vitality. *Journal of the American Society of Brewing Chemists* **57**(1), 18–23.

(15) Chilver, M.J., Harrison, J. and Webb, T.J.B. (1978) Use of immunofluorescence and viability stains quality control. *J. Am. Soc. Brew. Chem.* **36**, 13–18.

(16) Trevors, J.T., Merrick, R.L., Russell, I. and Stewart, G.G. (1983) A comparison of methods for assessing yeast viability. *Biotechnol. Lett.* **5**, 131–134.

(17) Mathieu, C., Van der Berg, L. and Iserentant, D. (1991) Prediction of yeast fermentation performance using the acidification power test. *Proceedings of the European Brewing Congress* **23**, 273–278.

(18) Quain, D.E. (1988) Studies on yeast physiology–impact on fermentation performance and product quality. *Journal of the Institute of Brewing* **95**, 315–323.

(19) Stewart, G.F. and Russell, I. (1993) Fermentation – the black box of the brewing process. *MBAA Tech. Quart.* **30**, 150–168.

7 Recent Developments in On- and Off-line Yeast Measurements Using Radiofrequency Impedance Methods

ROBERT TODD

Abstract   Over the last 10 years the radiofrequency impedance method of measuring yeast concentration has become accepted in many major breweries around the world. The method relies on detecting the capacitance of the yeast cell membranes, and gives a very rapid response to changes in yeast suspension concentration (< 1 second where necessary), and a linear response over a very wide concentration range. The Aber Instruments on-line Yeast Monitor has a long life probe that can easily be fitted to yeast mains, fermenters or storage vessels. An off-line laboratory version, the Lab Yeast Analyser, is also available.

An important feature of this method is that in dead yeast cells the membranes become electrically leaky, and therefore do not contribute significantly to the total capacitance detected. The derived concentration measurement therefore represents only live cells, i.e. those with intact membranes. Although the measurement of yeast concentration by this method effectively includes a viability correction, the current instruments cannot measure % viability as a separate variable. However, a method for making very rapid % viability estimation is currently being developed.

The main brewery applications are in automatic yeast pitching control, yeast recovery (including yeast inventory) and in yeast dosing for cask conditioned products. Developments are in progress which will extend the measurement range to lower yeast concentrations. The two approaches being applied to measuring lower concentrations are (a) the correction of a major source of impedance measurement error – electrode polarisation – by using a multifrequency measurement, and (b) for much lower concentrations, a new version of the Lab Yeast Analyser using a filtration technique. Currently we are experimenting with extracting more useful information from on-line yeast measurements, with the aim of providing an immediate warning of changes in the yeast slurry.

7.1 Introduction

Over the last 10 years the radiofrequency impedance method of measuring yeast concentration has become accepted in many major breweries around the world. The method relies on detecting the capacitance of the yeast cell membranes, and gives a very rapid response to changes in yeast suspension concentration (< 1 second where necessary), and a linear response over a very wide concentration range[1,2].

An important feature of this method is that in dead yeast cells the membranes become electrically leaky, and therefore do not contribute significantly to the total capacitance detected. The derived concentration measurement therefore represents only live cells, i.e. those with intact membranes, so a separate assessment of viability is often unnecessary.

The Aber Instruments on-line Yeast Monitor 320 (Fig. 7.1) has a long life probe which can easily be fitted to yeast mains, fermenters or storage vessels via a 25 mm port. An off-line laboratory version, the Lab Yeast Analyser, which has a 10 ml stirred sample chamber is also available[3].

Fig. 7.1   Aber Instruments 320 yeast monitor system, including the probe, headamplifier and accessories.

Currently, the main brewery applications are in automatic yeast pitching control, yeast recovery (including yeast inventory) and in yeast dosing for cask conditioned products. Figure 7.2 shows a typical pitching control system, which has, in many breweries, resulted in improved fermentation consistency[4]. Usually the control function is carried out by an existing brewery PLC; however, a self-contained measurement and control package suited to smaller, less automated breweries is also now available[5].

This chapter describes the present limitations of the technology and some of the new developments being undertaken at Aber Instruments in collaboration with colleagues at Aberystwyth University, and with the valuable assistance of our customers.

7.2   Limitations of the current technology and planned improvements

7.2.1   *Low concentration measurements*

The main limitation of the measuring technique is the increasing percentage errors at low biomass concentrations. We are measuring a bulk effect (unlike cell counters) so the capacitance representing the cells falls linearly with yeast concentration and eventually other small capacitance changes, not caused by the yeast itself, start to interfere. The three main sources of such variations are as follows.

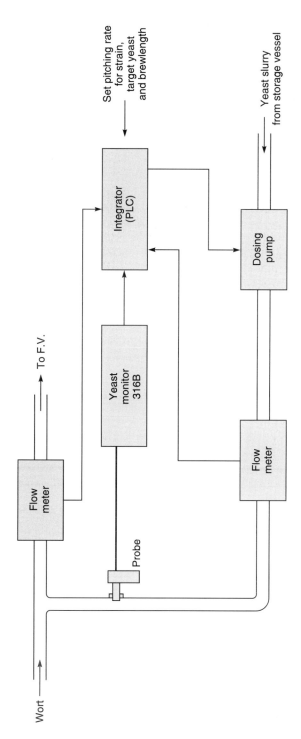

Fig. 7.2 Automatic yeast pitching system.

(1) *Crosstalk*   This is when changes in the *conductivity* of the yeast suspension cause artefactual changes in the measured *capacitance*. The conductivity varies due to ion concentration and mobility, and is very temperature sensitive. Large conductivity differences can occur between samples of the same yeast strain so it is important to minimise crosstalk error. There are two causes of crosstalk, the first is due to phase errors in the electronics, and the second is due to electrode polarisation. Electrode polarisation is the formation of a complex boundary region where the metal electrode meets the liquid. This region causes an additional capacitance component which adds to the true yeast capacitance and which, unfortunately, is dependent on conductivity and on the (rather variable) surface condition of the electrodes[6]. The upper curves in Fig. 7.3 show this effect when measuring a conductive medium with no cells. Note that the conductance range shown in the graph is much greater than that found in brewery applications; however, the effect is still significant when measuring low yeast concentrations.

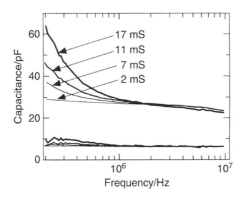

Fig. 7.3   Crosstalk and electron polarisation effects for cell-free media with conductances of 2, 7, 11 and 17 ms: upper curves without correction and lower curves with correction.

(2) *Temperature effects*   Changes in ambient temperature can cause some small errors in the electronics, but these are usually overshadowed by the effects of varying sample temperature. The water in the sample causes a background capacitance which is temperature dependent. Changes in the temperature of the sample cause changes in conductivity, which is important because of the crosstalk problem. The capacitance of the yeast cells itself is also temperature dependent, and the extent of the dependence varies between yeast strains.

(3) *Electrode contamination*   The main contamination problem found in brewing applications is trub (or sometimes yeast) sticking to the electrode pins. This causes small erratic capacitance changes as fouling builds up and then is dislodged again. The effect is completely negligible when measuring slurries, but is significant at fermenter yeast concentrations.

The effect of all these errors on the Lab Yeast Analyser accuracy is shown in Fig. 7.4. Note the important effect of cell size; the capacitance signal is much bigger for large yeasts. There are solutions to most of the above measurement problems and we

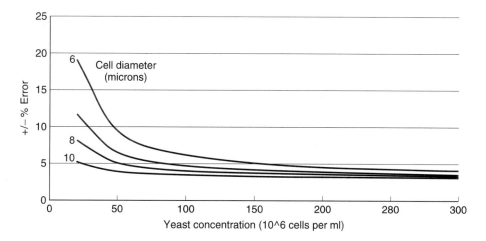

Fig. 7.4 Aber Instruments laboratory analyser 800 output showing typical percentage errors versus cell concentration for various sizes of yeast cell.

are developing a higher accuracy LYA incorporating several improvements. (i) *Improved electronics:* The electronic errors are reduced by a factor of about ten, and sample-temperature compensation or regulation is included. (ii) *Polarisation correction:* By measuring the capacitance at two frequencies, software can be used to correct for the electrode polarisation effect as shown in the lower curves of Fig. 7.3 and described fully in a recent paper[6]. (iii) *Improved electrodes:* By using ring electrodes which are flush with the side-wall of the sample chamber, sticking of trub and yeast can be minimised. Fig. 7.5 shows a prototype ring-electrode sample cell. We anticipate an improvement of about a factor of five in the low concentration accuracy, along with some relaxation of the measuring procedure requirements.

7.2.2 *In-fermenter measurements*

There is significant interest within the industry in monitoring viable yeast concentration on-line during the fermentation. Aber Instruments equipment is used routinely to do this by many pharmaceutical companies, but the fermenter conditions there are quite different; the fermenters are usually vigorously stirred and highly aerated. We have had mixed results so far in brewing fermenters. Clearly, the spatial distribution of yeast within the fermenter is likely to be uneven and time-varying, so measuring at a single point will not give such clear information as in the fully mixed fermenters. (Several probes can of course be multiplexed to one instrument to give a fuller picture). Also, the extremely slow fluid velocities in the brewing fermenter allow yeast and/or trub to settle on the measuring electrodes, creating random noise in the measurements. This effect is still important even when measurements are made in a recirculation loop.

Experiments have shown that the use of flush (rather than pin) electrodes, mounted with their faces vertically, gives a considerable improvement in the measurement noise. Figure 7.6 shows the progress of a fermentation (at Guinness, Dublin) using

Fig. 7.5 Prototype ring-electrode sample cell.

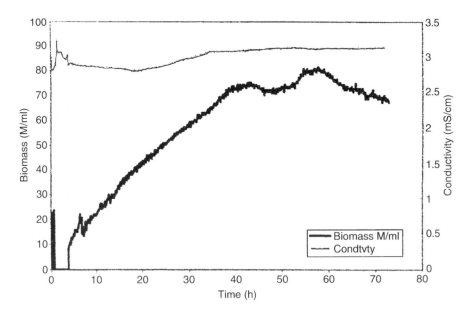

Fig. 7.6 Yeast concentration and conductivity measured on-line in a recirculating loop during a fermentation.

flush electrodes in a recirculation loop. New electrode probes combined with some of the other low concentration improvements described earlier will shortly be trialled in another brewery and should enable us to offer accurate in-fermenter monitoring within the next year.

7.2.3   *Viability*

Although the measurement of yeast concentration by the radiofrequency capacitance method effectively includes a viability correction, the current instruments cannot measure percentage viability as a separate variable. However, we are currently working on a method for making very rapid viability estimation, which does not use dyes and it is planned to incorporate this into an instrument similar to the Lab Yeast Analyser.

7.3   New low concentration measurement method

To extend the sensitivity of impedance-based yeast measurement significantly beyond even an improved Lab Yeast Analyser instrument, we have developed a new off-line measurement technique. This involves concentrating the yeast onto a filter membrane, and then measuring the RF impedance of the yeast on (and in) the membrane using a flat, porous electrode structure immediately behind the membrane. The electrode design concentrates the electric field into the yeast film. Figure 7.7 shows an early prototype apparatus. The membrane type and arrangement are selected to give

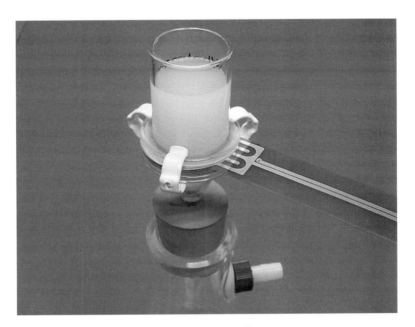

Fig. 7.7   Prototype apparatus for low-concentration yeast cell measurement.

minimum disturbance to the electric field and to give good measurement repeat-ability. A new membrane is used for each measurement. Vacuum is used to accelerate the filtration process.

In our prototype, measurement of a sample with around 10^6 cells/ml takes only a few minutes and gives a repeatability typically better than 10^5 cells/ml; the result, of course, is a measurement of viable cells only, and will ignore trub and dead cells. Figure 7.8 shows an experimental dilution curve, and Fig. 7.9 shows the measurement repeatability for three yeast concentrations. There are three measurements for each concentration, each made with a different membrane.

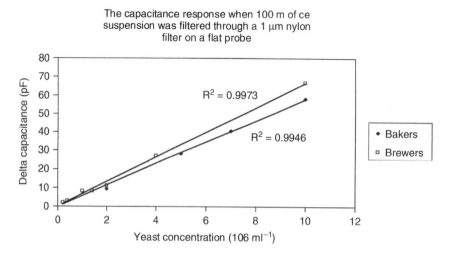

Fig. 7.8 Capacitance versus yeast concentration using the prototype filtration apparatus.

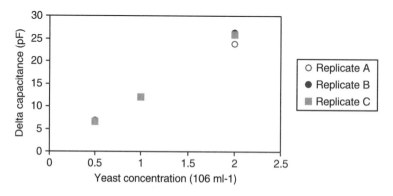

Fig. 7.9 Repeatability of measurements with the prototype filtration apparatus using a new filter membrane for each measurement, and three measurements at each of the concentrations.

7.4 Yeast condition monitoring

Our experience with yeast pitching systems over the last ten years has shown that, although the electronic measuring system is generally extremely stable (apart from the occasional probe failure) there are sometimes shifts in calibration, i.e. a change in the relationship between the capacitance-based concentration figure and that from another reference method, e.g. a viability corrected cell count. Fortunately, these shifts are usually quite small and often can be explained by events such as the switch from recovered to fresh yeast, for example. However, sometimes such calibration shifts occur without obvious reasons, and there is some evidence that they are associated with (and perhaps cause) changes in the fermentation process.

Our existing instruments measure the yeast impedance at a single frequency, optimised for brewing yeasts, and the reading is corrected for the influences of conductivity and of close cell packing to give a reliable viable yeast concentration value.

The impedance of a yeast suspension varies with frequency, and in the region of interest shows a relaxation curve (the beta-dispersion) in which the capacitance of the suspension falls with increasing frequency to reach a high frequency plateau. The beta-dispersion curves of yeast suspension capacitance versus frequency are shown in Fig. 7.10 for a range of cell concentrations. Measurements are usually made at a frequency close to the lower frequency plateau region. However, the beta-dispersion shifts along the frequency axis if the conductivity of the suspension changes and the fixed-frequency capacitance measurement is affected by this. Fortunately the error caused is accurately predictable. The calibration procedure of the instrument includes setting up the parameters required to correct for this effect. However, the single-frequency measurement cannot distinguish between a genuine viable concentration change and other changes in the yeast (e.g. cell size variation).

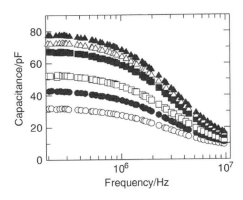

Fig. 7.10 Beta dispersion curves (after polarisation correction) for six different yeast concentrations.

Clearly the single-frequency measurement is missing a lot of information contained in the yeast impedance characteristic. As expected from theoretical modelling, it has been shown that certain changes in the yeast can be detected by making multi-frequency measurements. The exact shape of this curve and its position on the frequency axis depends on several physical properties of the yeast suspension[7].

The curve is described well by the expressions

$$\Delta C_f \propto PRC_m \frac{1 + (f/f_c)^{1-\alpha}\sin 90\alpha}{1 + 2(f/f_c)^{1-\alpha}\sin 90\alpha + (f/f_c)^{2-2\alpha}}$$

and

$$f_c = \frac{1}{RC_m(\frac{1}{S_i} + \frac{1}{2S_o})}$$

where ΔC_f is the capacitance above background at frequency f, P is the total volume fraction enclosed by cell membranes, f is the measuring frequency, f_c is the characteristic frequency, α is the Cole–Cole shape parameter, R is the cell radius, C_m is the membrane capacitance constant, S_i is the cytoplasm conductivity, and S_o is the suspending medium conductivity. A change in any of these parameters will result in a change in the curve. Note that the cell concentration acts as a scale multiplier without changing the shape of the curve as such. We can take account of the inevitable changes in the suspending-fluid conductivity (due to temperature, storage, acid wash, etc.), but any remaining change in the shape of the curve must be due to a change in cell size, cell membrane properties, cytoplasm conductivity, or a variation in the statistics of the (non-homogeneous) cell population.

It is practical to detect such changes (referenced to the yeast used to calibrate the instrument) on-line, and to provide an indication that *something* has changed other than simply viable cell concentration or conductivity. Although a specific diagnosis of

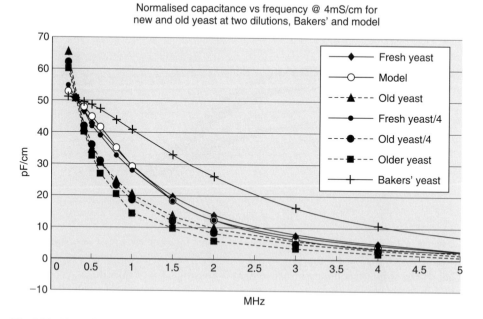

Fig. 7.11 Normalised capacitance versus frequency curves for baker's yeast and for one brewing yeast (three samples of different ages and viabilities, two of which were also diluted fourfold).

the fundamental cause of the impedance curve change cannot necessarily be made, the occurrence of a change could be used to prompt an investigation and possibly more specific lab tests.

The result of the curve shape assessment can be simplified by an algorithm in the instrument's microprocessor to give a single output number or 'yeast condition index'. This reflects any change away from the shape of the impedance curve established at calibration, but is substantially independent of yeast concentration and conductivity.

As this work is relatively new, we do not yet have on-line data, but the experimental data in Fig. 7.11 clearly show the differences in the curves produced from fresh brewing yeast, old low-viability yeast of the same strain, and yeast of a completely different type, a baking yeast. The data have been adjusted to remove the effects of concentration and conductivity changes. On-line trials of this technique are planned at a UK brewery in the near future.

7.5   Conclusions

Yeast monitoring by RF impedance measurement is well established in yeast pitching and yeast recovery and, with the larger yeast cells, can satisfactorily measure at fermenter concentration levels off-line. Improvements to the current technology will soon allow confident on- and off-line fermenter measurements. The filter membrane technique will extend the off-line measurement range to one or two orders of magnitude lower concentration than the direct suspension measurements. This opens up new applications in centrifuge and cask conditioning monitoring. By monitoring the shape of the yeast impedance curve with multi-frequency measurements, it has been shown that changes in the yeast or the presence of a different yeast strain can be detected. Radiofrequency impedance measurement provides a valuable tool which is adaptable to many different applications in the brewery.

Acknowledgements

Many thanks to Stephen Cunningham, John Yardley and Zoe Phillips for experimental data, and to Chris Davey, Doug Kell and Chris Boulton for ideas and many useful discussions.

References

(1)   Harris, C.M., Todd, R.W., Bungard, S.J., Lovitt, R.W., Morris, J.G. and Kell, D.B. (1987) The dielectric permittivity of microbial suspensions at radio frequencies; a novel method for the real-time estimation of microbial biomass. *Enzyme & Microbial Technology* **9**, 181–186 (general, simplified outline; demonstration of linearity of permittivity /capacitance with yeast biomass in cell suspensions and growing cultures; first description of Biomass Monitor approach).
(2)   Boulton, C.A., Maryan, P.S. and Loveridge, D. (1989) The application of a novel biomass sensor to the control of yeast pitching rate. *Proceedings of the 22nd European Brewing Convention*, Zürich, pp. 653–661 (uses of the Biomass Monitor to control yeast pitching).

(3) Carvell, J.P., Harding, C.L. and Oddi, L. (1998) Use of capacitance for off-line measurement of viable yeast concentration over a range of sample viabilities. Presented at The Institute of Brewing Asia Pacific Section, 25th Technical Convention, 22–27 March 1998, Perth, Western Australia.

(4) Carvell, J.P. (1997) Developments in on-line monitoring of viable yeast in the brewery process. *Ferment* **10**.

(5) Aber Instruments (1999) Yeast Pitching Control System (Model 430), Provisional Data Sheet. Aber Instruments Ltd, Aberystwyth.

(6) Yardley, J.E., Todd, R.W., Nicholson, D.J., Barrett, J., Kell, D.B. and Davey, C.L. (2000) Correction of the influence of baseline artefacts and electrode polarisation on dielectric spectra. *Journal of Bioelectrochemistry and Bioenergetics*, in press.

(7) Davey, C.L. (1993) *The Theory of the β-Dielectric Dispersion and Its Use in the Estimation of Cellular Biomass*. Aber Instruments Ltd., Aberystwyth (introductory book on the theory behind the Biomass Monitor).

8 Comparison of Yeast Growth Profiles of Fermentations Pitched on the Basis of Off-line Capacitance With Those Pitched Using Traditional Methods

STEPHEN CUNNINGHAM

Abstract The problems associated with under and over pitching brewery fermentations have been widely reported. There is therefore a requirement for accurate yeast pitching rates to ensure that fermentations are predictable and consistent. The traditional methods currently used by many breweries rely on the measurement of cell mass or cell numbers. These methods require correction for yeast viability, usually using viability stains such as methylene blue. There are widespread concerns about the accuracy of methylene blue when yeast viability is less than 90%, but the method is also prone to human error and subjectivity of the operator. The use of cellular capacitance to determine the viable cell number in pitching yeast slurries is now widespread and can be performed both on- and off-line. As only viable cells with intact cell membranes produce a signal, dead cells and trub do not interfere with the readings. The capacitance technique produces results in a matter of seconds, eliminating the need for time-consuming dilutions, and also the inaccuracies due to viability staining and human error. In this set of experiments wort fermentations were set up using a number of brewing yeast strains. Yeast growth was also monitored during fermentation using off-line capacitance and manual cell counts. The results obtained showed that the fermentions pitched on the basis of off-line capacitance gave more consistent fermentation profiles than those pitched on the basis of wet weight. As the capacitance technique measured the total viable yeast biovolume in the fermenter samples a different profile from that measured by manual cell counting was sometimes observed, although this can be explained by changes in yeast cell size during the fermentation.

8.1 Introduction

The non-conducting nature of the yeast plasma membrane allows a build of charge when the cells are placed in an electric field[1]. The resulting capacitance can be measured, and it is directly proportional to the concentration of viable cells. Non-viable cells will possess 'leaky' membranes allowing charged particles to pass through; these cells cannot be polarised and do not contribute to the capacitance signal. The on-line capacitance measurement of viable yeast concentration is used in many breweries for automatic yeast pitching, and has been shown to reduce the variability of pitching rates associated with standard laboratory methods[2,3].

The traditional laboratory methods for determining viable yeast concentration rely on measuring cell mass or cell numbers. Cell mass is determined either by filtration or centrifugation, and inaccuracies in these results may occur due to varying levels of trub, other particulate matter and cell packing density. Generally cell numbers are measured by a manual cell counting procedure using a haemocytometer or a Coulter counter. Both methods rely on the yeast slurry being carefully diluted, which can be time consuming and may lead to inaccuracies in the results. These procedures require the results to be adjusted for yeast viability usually using methylene blue. There are widespread concerns about the accuracy of methylene blue when yeast viability is less than 90%, and there could be an overestimation of the viable cell concentration in these samples. This method is also prone to human error and the subjectivity of the operator[4]. Thus an off-line instrument has been developed that uses the same capacitance principle as the on-line yeast monitors used successfully in a number of breweries. The off-line instrument is advantageous as it can be used for a wider range of samples. The results can be obtained in seconds reducing the laboratory time required for yeast analysis, but also reducing inter- and intra-operator errors. The data presented illustrate the flexibility of the instrument in that it can be used for determining the viable yeast concentrations in slurries up to 70% spun solids, but can also be used to follow fermentations in both laboratory and production scale.

8.2   Materials and methods

The yeast strain used in this study was a production brewing ale strain of *Saccharomyces cerevisiae*. Yeast suspensions were prepared comprising similar yeast concentrations, but had varying levels of trub. Viable yeast concentration was determined using the off-line capacitance method (Lab Yeast Analyser [LYA], Aber Instruments Ltd, UK). The brewery fermentations were monitored using the LYA, or manual counts using a haemocytometer or Coulter counter, and adjusted for viability using methylene blue. The data were obtained from either a yeast propagator or industrial 200hl brewery fermentations. A set of laboratory fermentations in 2 l EBC tube fermenters was also set up. These were pitched with a brewing lager strain of *S. cerevisiae* using the results obtained from the Lab Yeast Analyser or manual cell counting. The fermentations were monitored using the LYA.

8.3   Results

Wort can contain varying amounts of particulate matter or trub. Trub can interfere with yeast measurements when determining cell mass by filtration or centrifugation. The data presented in Fig. 8.1 illustrate that the trub content of yeast slurry does not produce a capacitance reading. When the trub was added in increments to a standard yeast suspension containing approximately 3% spun solids of viable yeast no effect on the capacitance signal was observed, even at three times higher than normal levels.

   The growth of yeast within a brewery propagator (Fig. 8.2) was monitored by manual counts using a haemocytometer and using a Lab Yeast Analyser (LYA). The reading from the LYA started to increase after 20 h, whereas cell numbers as measured using a haemocytometer did not significantly increase until 26 h after inoculation. Both readings increased at a similar rate after 26 h, although the reading from the LYA was always higher. Although the readings were dissimilar the LYA and the haemocytometer counts did follow a similar trend.

   A number of brewery ale fermentations were followed by manual cell counting

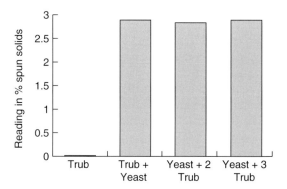

Fig. 8.1   Effect of wort trub on the capacitance signal from a yeast slurry.

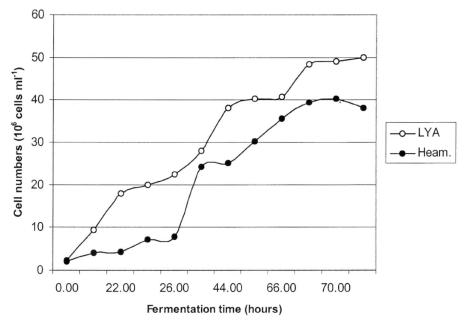

Fig. 8.2   A propagator fermentation followed using the Lab Yeast Analyser (○) and by manual cell counting (●).

using a haemocytometer, Coulter counter and the Lab Yeast Analyser (Fig. 8.3). All the fermentations were pitched at 30 million cells/ml and were completed after 72 h of fermentation. The data showed that brewery fermentation can be easily followed using the Lab Yeast Analyser, with the results being closely correlated to the manual cell counting method. The results from the Coulter counter were more variable and did not correlate well with the data obtained from the manual counting or the Lab Yeast Analyser. Laboratory scale fermentations were pitched on the basis of the LYA reading (Fig. 8.4) and were monitored by manual cell counting or using the LYA. The

(a)

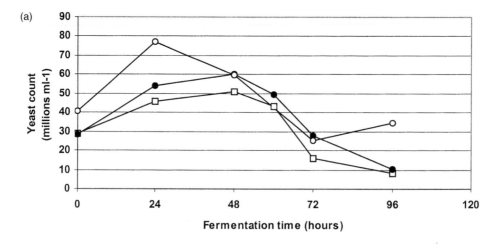

(b)

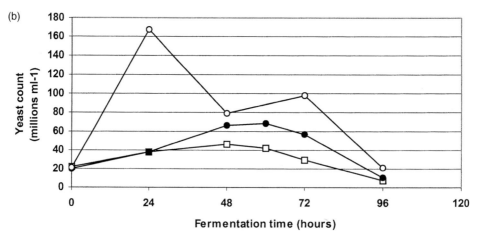

(c)

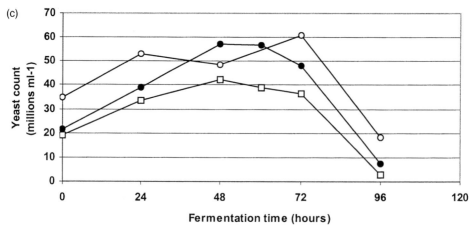

Fig. 8.3 Yeast cell counts during brewery fermentations determined by Coulter counter (○), manual cell counts (●) and using the Lab Yeast Analyser (□).

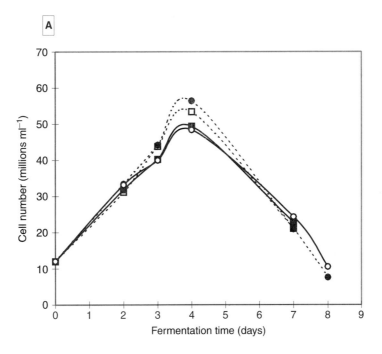

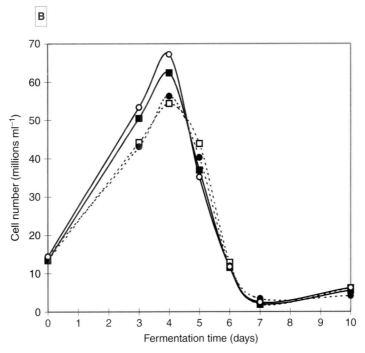

Fig. 8.4 Duplicate laboratory scale fermentations with two lager yeast strains (full lines and dashed lines, respectively) monitored using the Lab Yeast Analyser (circles) and by manual cell counting using a haemocytometer (squares).

results obtained from the LYA correlated well with data obtained from manual cell counts adjusted for viability using methylene blue.

8.4   Discussion

The LYA did exhibit an increase in cell numbers over the first 26 h of the fermentation, whereas cell numbers as counted manually remained relatively significantly lower. This can be accounted for by the fact that the LYA does not measure individual cells, but will measure the biovolume of the yeast within a sample. In the first 24 h after inoculation the cell numbers do not change significantly, but the yeast will be going through various physiological changes including an increase in cell size. As the capacitance readings are directly related to cell volume, an increase in cell size would have resulted in an increased capacitance signal and a corresponding increase in the observed cell number. The procedure for cell counting does not generally take into account the buds on vegetative cells. The buds are also enclosed by a plasma membrane and they are taken into account by the LYA measurement.

The results obtained from laboratory and brewery fermentation showed that there was a good correlation between manual cell counts and the readings from the LYA. The data obtained from the Coulter counter during brewery fermentations was more variable and did not always correlate with the results from the other methods. The variability may have been due to the sample dilutions that are necessary for this instrument. A previous report has shown that the manual counting method is very reproducible if the same individual performs each count. Errors can occur when different individuals count each of the samples, whereas the LYA was very reproducible over a wide range of sample concentrations with no inter-operator errors[5].

Previous data have shown that laboratory scale fermentations that were pitched on the basis of the LYA were also more consistent than those pitched using the traditional methods, i.e. centrifugation and optical density methods[2]. The results from traditional methods can be affected by the trub content of the yeast slurry. The LYA reading is not affected by the variation in the amount of trub present as it does not interfere with the capacitance signal (Fig. 8.1). Variation can also occur due to the use of methylene blue, which has been reported to be inaccurate when the yeast viability is below 90%[4].

The detection limit of the LYA is dependent upon cell size, as larger yeast cells (greater than 7 μm in diameter) produce a higher capacitance signal and can be detected at concentrations of 15 million/ml and above. Generally 30 to 50 million cells/ml is the lower operating range for smaller yeast cells (i.e. less than 6 μm in diameter). The data presented illustrated that the Lab Yeast Analyser can be used to measure the viable yeast concentration over a wide range of brewery samples including laboratory scale, propagator and fermenter samples to yeast slurries. The results are obtained much more rapidly than the traditional laboratory methods, and with increased accuracy due to eliminating the time consuming serial dilutions and the requirement for methylene blue staining.

Acknowledgements

Aber Instruments Ltd wished to thank Kathy van der Speigle from Interbrew, Leuven and Arno Mathee from South African Breweries Rosslyn Brewery for supplying the data for this manuscript.

References

(1) Harris, C.M., Todd, R.W., Bungard, S.J., Lovitt, R.W., Morris, J.G. and Kell, D.B. (1987) *Enzyme Microbial. Technology* **9** 181–189.

(2) Boulton, C.A., Maryan, P.S. and Loveridge, D. (1989) The application of a novel biomass sensor to the control of yeast pitching rate. *Proceedings of the European Brewery Convention Congress* **22**, 653–661.

(3) Maca, H.W., Barney, M., Goetzke, G., Daniels, D. and Ryder, D. (1994). The use of radiofrequency capacitance for the measurement of yeast viable biomass and its use in the automatic pitching of fermentations. *MBAA Tech. Quart.* **31**, 146–148.

(4) O'Connor-Cox, E., Mochaba, F.M., Lodolo, E.J., Majara, M. and Axcell, B. (1997) Methylene blue staining at your own risk. MBAA *Tech. Quart.* **34**, 306–312.

(5) Moore, J. and Storey, K. (1998) *Evaluation of the Aber Instruments Lab Yeast Analyser.* Evaluation Report, Brewing Research International.

(6) Carvell, J.P., van de Speigle, K. and Matthee, A. (2000) *Journal of the American Society of Brewing Chemists*, in press.

Part 2   The Life and Death of the Yeast Cell

9 Yeast Growth and Nutrition

JOHN HAMMOND

Abstract   When yeast cells are pitched into brewery wort there is a lag phase during which little appears to happen. However, within the yeast cells many changes are taking place. Initially endogenous reserves are utilised while the cells adapt to use the nutrients in the wort. Thereafter rapid nutrient assimilation and growth occur. In the early stages of fermentation, oxygen present in the wort is used for the synthesis of membrane components essential for growth. Once the oxygen is depleted cell division and growth become restricted, the yeast cells become wholly fermentative and the specific gravity of the wort begins to decline rapidly.

Although the majority of alcohol formation is carried out by non-dividing cells, growth of new cells is essential for the maintenance of a healthy and vigorous yeast population during the fermentation phase. The nutrients required for this are obtained solely from the wort. Like any living organism, yeast requires sources of carbon, nitrogen, trace elements and vitamins. Fermentable sugars in brewery worts provide the carbon whilst amino acids and peptides provide the nitrogenous components. Vitamins derived from the malt are essential for enzyme function, as are trace elements comprising of metal ions, sulphate and phosphate.

For beer production the supply of oxygen is restricted since the brewer has no desire to grow yeast at the expense of alcohol production. This is not the case when a new yeast culture is being propagated for introduction into the brewery. Here yeast growth is required, and a number of strategies have been developed for this purpose. They all involve heavy aeration and include batch, fed-batch and semi-continuous processes.

With a knowledge of the nutrient needs of growing and fermenting yeast cells it is possible to control the fermentation and propagation processes in order to achieve the desired outcome. However, it must be emphasised that at present control is by no means perfect, reflecting our still limited knowledge of yeast nutrition and metabolism under brewing conditions.

9.1   Introduction

The production of beer is one of the oldest biotechnologies known to man, and since early times the purpose of a brewery fermentation has remained unchanged: the conversion of sweet, viscous wort into alcoholic, refreshing beer. This is brought about by the activity of brewer's yeast, which not only converts wort sugars to carbon dioxide and ethanol but also produces the other by-products of fermentation so essential to beer flavour. Although this process is largely fermentative, the growth of fresh yeast cells is an essential part of the process; without growth only poor quality beer can be made.

9.2   Yeast growth

When yeast cells are pitched into brewery wort there is a lag phase during which very little appears to happen. Only after several hours does the specific gravity of the wort begin to decline as sugars are taken up and metabolised by the yeast. Before this occurs the yeast must use an endogenous carbon source, glycogen. This is accumulated during the later anaerobic stages of fermentation and can represent up to 40% of the dry weight of the cells. It is rapidly dissimilated when yeast is pitched into aerated wort.

The oxygen present early in fermentation is rapidly used up for the synthesis of membrane components essential for growth. Once the oxygen is depleted, cell division slows and the yeast cells switch to anaerobic fermentation, and the rapid decline in specific gravity begins. This process is highly temperature dependent, faster sugar utilisation occurring at ale fermentation temperatures (16 to 23°C) than at lager fermentation temperatures (8 to 15°C). The yeast strain used is also important, lager yeasts being capable of growth and fermentation at lower temperatures than ale yeasts while the latter have higher maximum growth temperatures.

Growth of new cells is essential for the maintenance of a healthy and vigorous yeast population during the fermentation phase, and the nutrients required for this are obtained from the wort. Yeast requires sources of carbon, nitrogen, trace elements, vitamins and lipids. Fermentable sugars in brewery wort provide the carbon while amino acids and peptides provide the nitrogenous components. Vitamins derived from the malt are essential for enzyme function, as are trace elements comprising of metal ions, sulphate and phosphate. Lipids can be synthesised provided that there is a supply of oxygen.

9.3 Carbohydrates

Of the sugars present in wort, brewer's yeasts can utilise fructose, glucose, maltose, sucrose and maltotriose. In addition, most can use isomaltose and some can use panose and maltotetraose. None can utilise higher oligomers of glucose, the dextrins. Many yeasts remove the sugars from wort in a strict order; sucrose, broken down extracellularly by invertase, is the first to go, followed by the monosaccharides glucose and fructose, and, finally, by maltose and maltotriose. The monosaccharides are taken up by facilitated diffusion involving common membrane carriers. Both high and low affinity systems appear to exist, a number of genes, including *SNF3* and *HXT1-4*, being involved[1]. Maltose and maltotriose are actively taken up by specific permeases.

The most important factor determining the rate of fermentation of wort is the rate at which the most abundant sugar, maltose, is utilised. Maltose fermentation requires the products of at least one of five *MAL* loci (*MAL1-4* and *MAL6*) each consisting of three genes; *MAL S* (-glucosidase), *MAL T* (permease) and *MAL R* (activator which regulates the other genes). Different yeast strains have different combinations of *MAL* genes. They all map at or near telomeres of chromosomes, suggesting that they were made by duplication and rearrangement. The expression of *MAL S* and *MAL T* is regulated by induction by maltose and repression by glucose. This control involves both *MAL R* and *MIG1*[2]. When the glucose concentration is high (> 1% w/v) the *MAL* genes are repressed and so in brewery fermentations maltose uptake normally occurs only when 50% of glucose has been taken up by the yeast.

Spontaneously arising mutants, resistant to the repressive effects of glucose, can be generated using the glucose analogue 2-deoxy-D-glucose, and de-repressed mutants of ale yeasts have been produced in this way[3]. One recent improvement to a brewing yeast has involved genetic modification to increase the gene dosage of constitutively expressed *MAL T* which resulted in significantly faster fermentation rates[4]. In a similar way repression by *MIG1* has also been overcome by disruption of the gene[2].

Control of maltose utilisation occurs at three levels (see Fig. 9.1).

(1) Maltose induces and glucose represses transcription of *MAL S* and *MAL T*. This is brought about by the constitutively expressed MalRp regulatory protein binding near the promoters of these two genes.

(2) Post transcriptional control – Addition of glucose to induced cells causes up to a 70% increase in lability of mRNA containing the *MAL S* transcript. This suggests that there are factors binding to mRNA in the absence of glucose which prevent degradation.

(3) Post translational modification – In the presence of glucose, maltose permease is either reversibly converted to a form with decreased affinity or is degraded (glucose inactivation).

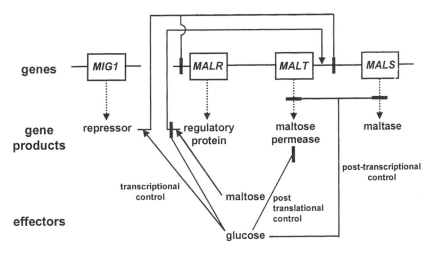

Fig. 9.1 Genetics of maltose metabolism (arrows represent induction and bars represent repression).

In addition the *MIG1* gene has been shown to encode a repressor protein which represses all three *MAL* genes. This activity is induced by glucose.

Sucrose is broken down extracellularly by invertase to glucose and fructose, which are then taken up normally. However, since brewer's yeasts utilise fructose and glucose at different rates this can have flavour implications where large amounts of sucrose adjuncts are used. Since yeast consumes fructose more slowly, a portion of this very sweet sugar can remain at the end of fermentation and affect the taste of the beer produced.

Sucrose degradation is controlled by six genes scattered throughout the genome, each of which encodes invertase (*SUC1-5* and *SUC7*)[5]. Most yeast strains do not carry *SUC* genes at all locations although *SUC2* always appears to be present in laboratory strains. In much the same way as the *MAL* genes, all other *SUC* genes are near telomeres and probably have been derived by telomeric rearrangement.

Once sugars have entered the yeast cell they are all metabolised in the same way via phosphorylated intermediates. The di- and trisaccharides are converted to hexose monomers and, together with glucose and fructose, are largely metabolised by the Embden–Meyerhof–Parnas pathway.

9.4   Nitrogen sources

Wort nitrogen levels have a marked effect on yeast growth. Below about 100 mg/l free amino nitrogen, growth is nitrogen dependent. Above this value growth becomes less dependent and free amino nitrogen levels above about 220 mg/l have little effect.

Brewer's yeasts can grow on ammonium salts as their sole nitrogen source but, in wort, their requirements for nitrogenous nutrients are supplied by amino acids and low molecular weight peptides. As a result of interactions between the various amino-acid transport systems present in yeast, amino acids are removed from wort in an orderly fashion[6]. Four groups of amino acids have been identified in ale fermenta-tions: group A (arginine, lysine, aspartate, asparagine, glutamate, glutamine, serine, threonine; absorbed immediately), group B (leucine, isoleucine, valine, methionine, histidine; absorbed gradually), group C (alanine, glycine, phenylalanine, tryptophan, tyrosine, ammonia; absorbed after a lag) and group D (proline; hardly absorbed). In lager fermentations, the order of uptake is similar, although arginine, aspartate and glutamate tend to be taken up more slowly[7]. The general amino acid permease (GAP) appears not be synthesised early in brewery fermentations because of the presence of ammonia in wort. As a result, amino acids in groups A and B are transported into yeast by specific permeases. Only when ammonia has been used are the group C amino acids taken up by the GAP.

Calculation of the rates of absorption of α-amino nitrogen and total nitrogen from wort indicates that absorption of other nitrogenous compounds must accompany that of amino acids[7]. These other compounds are small peptides, about 40% of which are removed from wort during fermentation.

9.5   Vitamins

The requirement for growth factors by yeast is very strain-dependent. Most brewer's yeasts have an absolute requirement for biotin, which is used by carboxylase enzymes and is essential for lipid synthesis. Many strains also need pantothenate although some can overproduce this growth factor and excrete it into the beer. This growth factor is a component of coenzyme A and so is needed for carbohydrate and lipid metabolism as well as being involved in the synthesis of sulphur containing amino acids. Inositol, a component of phospholipids, is sometimes required, whereas pyri-doxine and thiamine seem only to be needed by brewer's top yeasts. They are, respectively, vital components of enzymes involved in carbohydrate fermentation and amino acid metabolism. Other growth factors seem not to be necessary for successful fermentation.

Brewer's wort is a rich source of growth factors, and it is unusual for deficiencies to occur, although fermentation defects due to shortages of biotin and inositol have been reported.

9.6 Trace elements

Brewer's yeasts have a wide requirement for minerals. Among the cations required are zinc, manganese, magnesium, calcium, copper, potassium and iron. Zinc, manganese, iron, copper and magnesium are all enzyme cofactors either as a component of the catalytic centre, as an activator or as a stabiliser. Potassium (and hydrogen (protons)) are components of the transport systems for nutrient uptake. Calcium stimulates growth but its main role is in yeast flocculation.

It is rare for brewery worts to be deficient in these cations, with the exception of zinc, which can often be present in suboptimal concentrations[8]. If concentrations of zinc ions drop below 0.1 mg/l, slow fermentations result, which can be restored to normality by addition of zinc salts. If concentrations of zinc ions become excessive (over 0.6 mg/l), yeast growth is depressed unless concentrations of manganese ions are similarly high[9].

In addition to cations, brewer's yeasts also require larger concentrations of phosphate and sulphate for synthesis of organic phosphorus- and sulphur-containing compounds.

9.7 Oxygen and lipid metabolism

Brewer's yeast has an absolute requirement for oxygen for both sterol and unsaturated fatty-acid synthesis. Sterols and unsaturated fatty acyl residues are both vital for the structure and function of the yeast plasma membrane. Yeast harvested at the end of a brewery fermentation typically contains 0.1% (w/w) sterol and 1% (w/w) of the sterol precursor squalene. On pitching into aerated worts, the sterol content of the cell increases 10-fold. Once the oxygen present in wort has been consumed, the sterol content of yeast cells decreases, as the cells grow and divide, until eventually a limiting level (0.1%, w/w) is reached and reproductive growth ceases[10].

The requirement for wort oxygenation for satisfactory fermentation can be eliminated by addition of ergosterol and oleic acid to the wort[11] or by addition of lipids extracted from spent grains[12]. Addition of spent-grain lipids or use of aerobically propagated yeast, rich in lipids, can shorten fermentation times, although this is at the expense of greater yeast crops which decrease the overall efficiency of the conversion of sugar to alcohol.

The amount of oxygen required for sterol synthesis and satisfactory fermentation differs widely between different strains of brewer's yeasts. The oxygen requirements change dramatically if oxygen is supplied in increments. For instance, strain NCYC 1062 normally requires an oxygen concentration of more than 40 mg/l for satisfactory fermentation, but this can be decreased to only 4 mg/l if the oxygen is provided incrementally over 12 h. It may be that such strains require oxygen later in a fermentation than low oxygen-requiring strains and so, when oxygen is added only once, a high initial content in wort is required to ensure the presence of some oxygen at a later stage[13].

In a typical fermentation it can be calculated that only about 30% of the available oxygen is used for lipid synthesis. Since the sugar content is too high for conventional aerobic respiration to occur it appears that yeast must 'waste' oxygen in some way. There is growing evidence for suggesting that respiration and a functioning electron transport chain could be significant in brewery fermentations, especially since it is known that functional mitochondria are required[14].

At first sight it would seem sensible for brewers to use non-growing cells for fermentation. In theory this would maximise the conversion of sugar to alcohol since no carbohydrate would be converted to biomass. However, the specific rate of sugar utilisation is directly related to the specific growth rate of the yeast in the fermenter[15]. Throughout fermentation the specific rate of sugar utilisation declines as the specific growth rate drops. When yeast-mass production ceases, maintenance of the yeast population requires only a low rate of sugar utilisation. For this reason, non-growing cells cannot be used for fermentation unless they are present in very high concentrations[13], either immobilised or in some form of continuous fermenter. However, such systems, although efficient alcohol producers, do not produce good quality beer unless yeast growth is permitted. As described above, during batch fermentations yeast control mechanisms determine the sequential uptake of amino acids and carbohydrates and the biosynthesis of lipids for regulated cell growth and the maintenance of cell integrity. In particular amino-acid metabolism is very important for beer quality, being linked not only to yeast growth but also to the production of vicinal diketones, sulphur compounds, fusel alcohols and esters, all important determinants of beer flavour. Reactors containing high concentrations of yeast and in which little growth occurs show severe limitations with respect to amino-acid uptake and transfer of oxygen to membrane lipid synthesis. This results in abnormal levels of higher alcohols and esters in the beer. Consequently, all successful continuous reactors described to date have either involved two stages (growth and fermentation) (see Fig. 9.2)[16] or have ensured controlled oxygenation within the reactor for continuous growth and cell release[17].

9.8   Yeast propagation

One area within a brewery where yeast growth has always been required is the yeast propagator. Routine replacement of yeast cultures after a few fermentations is nowadays considered good practice in order to avoid build-up of variants or contaminant organisms in the pitching yeast. Typically, cultures are grown from laboratory stock cultures by serial transfer into increasing volumes of hopped wort until 20 to 50 litres of yeast culture is obtained. The laboratory-grown culture is then transferred to one or more propagation vessels where yeast growth is encouraged by heavy aeration. Propagation schemes vary from brewery to brewery but, in all cases, large dilutions on transfer are avoided to minimise the chances of contamination. The

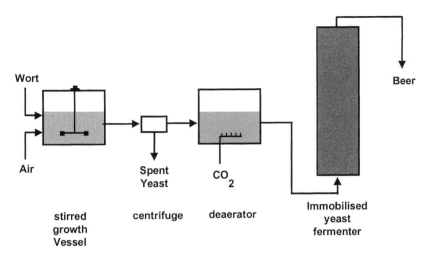

Fig. 9.2 Kirin two-stage continuous fermenter.

laboratory-based part of a yeast propagation is usually carried out at 23 to 28°C, but the temperature is lowered in the yeast-propagation plant to be closer to that employed for beer fermentation. In principle, however, such systems are essentially brewery fermentations with added aeration and, because of the high levels of sugar present in the wort, yeast yields are not much higher than those found in beer production. The use of fed batch fermenters, where the sugar concentration is kept low by incremental feeding, can increase yields fourfold when compared with conventional propagation systems. Such systems are yet to find favour in commercial practice.

9.9 Brewery worts

Well oxygenated all-malt wort provides all the requirements for yeast growth. Problems are usually experienced only when high levels of adjuncts are used because of the dilution of nitrogen, vitamins and ions. In these circumstances it may be necessary to supplement the wort with nutrients in the form of yeast foods. A special case of this is high gravity brewing, where the additional fermentable material is usually supplied as sugar syrups. Nutrition is especially important in such cases since oxygen levels may be low (the solubility of oxygen is lower in high gravity wort than in conventional wort), there may be a very high carbon:nitrogen ratio in the wort and the levels of growth factors and ions may be limiting. Additionally, towards the end of fermentation the yeast will experience an additional stress in the form of high levels of alcohol.

9.10 Alcohol tolerance

Ethanol inhibits both growth and fermentation in a non-competitive fashion, growth being more sensitive than fermentation[18]. In general, ale yeasts are less tolerant than lager yeasts, although growth conditions can markedly affect this, yeasts grown aerobically or in the presence of unsaturated fatty acids being less alcohol sensitive. Increasing temperatures increase the toxic effects of alcohol[19], and ethanol decreases the maximum temperature and increases the minimum temperature for yeast growth. Ethanol also affects cell viability, and poor yeast viabilities in high gravity fermentations have been attributed to this.

Ethanol is a non-competitive inhibitor of the transport of glucose, maltose, ammonia and several amino acids. Yeasts with membranes enriched in unsaturated fatty acids are more able to withstand high alcohol concentrations than cells not so enriched. It is likely that one of the effects of ethanol is to enhance the passive influx of hydrogen ions across the plasma membrane together with a direct inhibition of the plasma- membrane ATPase, both of which would dissipate the proton gradient across the membrane and so inhibit solute transport. In addition, from studies on the effects of ethanol on thermal death, it is clear that mitochondria also play an important part in ethanol tolerance.

The problems of poor yeast viability and stuck fermentations often encountered in very high gravity fermentations can be largely overcome by supplying extra nitrogenous compounds, sterols and unsaturated fatty acids[19]. Supplementation of fermentations in this way enables beers to be produced containing more than 16% (v/v) ethanol, but at the expense of additional yeast growth. Addition of extra magnesium and calcium ions has also been shown to increase ethanol yields in high gravity fermentations[20]. Thus, it seems that the normally observed alcohol-tolerance limits in brewery fermentations are not a property of the yeast strain but are determined by fermentation conditions. Indeed, using a sake-like fed-batch fermentation system, brewer's yeasts have been reported to produce a 'beer' containing more than 25% (v/ v) alcohol[21]. The importance of good yeast growth in such conditions cannot be overstressed.

9.11   Conclusions

With a knowledge of the nutrient needs of growing and fermenting cells, it is possible to control both propagation and fermentation. However, at present, our control of these processes is by no means perfect, reflecting our limited knowledge of yeast nutrition and metabolism under brewery conditions. It is clear, however, that a growing yeast cell is a healthy cell and healthy cells are required for successful fermentation.

References

(1)   Entian, K-D. (1996) Plasma membrane activity and the uptake of sugar and amino acids. *Belgian Journal of Brewing and Biotechnology* **21**, 43–47.
(2)   Klein, C.J.L, Olsson, L., Ronnow, B., Mikkelsen, J.D. and Nielsen, J. (1996) Alleviation of glucose repression of maltose metabolism by *MIG1* disruption in *Saccharomyces cerevisiae*. *Applied and Environmental Microbiology* **62**, 4441–4449.
(3)   Jones, R.M., Russell, I. and Stewart, G.G. (1986) The use of catabolite derepression as a means of

improving the fermentation rate of brewing yeast strains. *Journal of the American Society of Brewing Chemists* **44**, 161–166.

(4) Kodama, Y., Fukui, N., Ashikari, T., Shibano, Y., Morioka-Fujimoto, K., Hiraki, Y. and Nakatani, K. (1995). Improvement of maltose fermentation efficiency: constitutive expression of *MAL* genes in brewing yeast. *Journal of the American Society of Brewing Chemists* **53**, 24–29.

(5) Carlson, M. (1987) Regulation of sugar utilisation in *Saccharomyces* species. *Journal of Bacteriology* **169**, 4873–4877.

(6) Pierce, J.S. (1982) Amino acids in malting and brewing. *Journal of the Institute of Brewing* **88**, 228–233.

(7) Palmqvist, U. and Ayrapaa, T. (1969) Uptake of amino acids in bottom fermentations. *Journal of the Institute of Brewing* **75**, 181–190.

(8) Jacobsen, T., Lie, S. and Hage, T. (1981) Wort quality and the zinc content of malt. *Proceedings of the European Brewery Convention Congress* **18**, 97–104.

(9) Helin, T.R.M. and Slaughter, J.C. (1977) Minimum requirements for zinc and manganese in brewer's wort. *Journal of the Institute of Brewing* **83**, 17–19.

(10) Aries, V. and Kirsop, B.H. (1977) Sterol synthesis in relation to growth and fermentation by brewing yeasts inoculated at different concentrations. *Journal of the Institute of Brewing* **83**, 220–223.

(11) David, M.H. and Kirsop, B.H. (1972) The varied response of brewing yeasts to oxygen and sterol treatment. *Proceedings of the American Society of Brewing Chemists* **30**, 14–16.

(12) Taylor, G.T., Thurston, P.A. and Kirsop, B.H. (1979) The influence of lipids derived from malt spent grains on yeast metabolism and fermentation. *Journal of the Institute of Brewing* **85**, 219–227.

(13) Kirsop, B.H. (1982) Developments in beer fermentation. *Topics in Enzymology, Fermentation and Biotechnology* **6**, 79–131.

(14) O'Connor-Cox, E.S.C., Lodolo, E.J. and Axcell, B.C. (1996) Mitochondrial relevance to yeast fermentative performance: a review. *Journal of the Institute of Brewing* **102**, 19–25.

(15) Searle, B.A. and Kirsop, B.H. (1979) Sugar utilisation by a brewing yeast in relation to the growth and maintenance phases of metabolism. *Journal of the Institute of Brewing* **85**, 342–345.

(16) Inoue, T. (1995) Development of a two-stage immobilised yeast fermentation system for continuous beer brewing. *Proceedings of the European Brewery Convention Congress* **25**, 25–36.

(17) Mensour, N., Margaritis, A., Briens, C.L., Pilkington, H. and Russell, I. (1995) 'Gas lift' systems for immobilised cell fermentations. *Proceedings of the European Brewery Convention Symposium: Immobilised yeast Applications in the Brewing Industry* **24**, 125–133.

(18) Kalmokoff, M.L. and Ingledew, W.M. (1985) Evaluation of ethanol tolerance in selected *Saccharomyces* strains. *Journal of the American Society of Brewing Chemists* **43**, 189–196.

(19) Casey, G.P. and Ingledew, W.M. (1985) Re-evaluation of alcohol synthesis and tolerance in brewers yeast. *Journal of the American Society of Brewing Chemists* **43**, 75–83.

(20) Stewart, G.G., D'Amore, T., Panchal, C.J. and Russell, I. (1988) Factors that influence the ethanol tolerance of brewer's yeast strains during high gravity wort fermentations. *Technical Quarterly of the Master Brewers Association of the Americas* **25**, 47–53.

(21) Casey, G.P. and Ingledew, W.M. (1986) Ethanol tolerance in yeasts. *CRC Critical Reviews in Microbiology* **13**, 219–279.

10 Role of Metal Ions in Brewing Yeast Fermentation Performance

GRAEME WALKER

Abstract   The nature and concentration of metal ions in brewing liquor have long been recognised as important determinants of final beer flavour. In relation to brewing yeast and fermentation, most interest to date has focused on the roles of zinc and calcium ions in influencing wort attenuation and yeast flocculation, respectively. Until relatively recently, little attention has been paid to the role of magnesium ions in brewing yeast physiology and fermentation performance. Magnesium is absolutely essential for yeast growth and metabolism, and the bioavailability of this cation in malt wort is now recognised as being very important for efficient brewing fermentations. This paper reviews recent work which has revealed key roles for magnesium in dictating fermentative metabolism of brewing strains of *Saccharomyces cerevisiae* and in governing the response of brewing yeasts to environmental stress. Specifically, evidence is presented which indicates that brewer's wort, and brewer's yeast, may not be fully optimised with respect to magnesium status for efficient fermentative conversion of wort sugars to ethanol. Results also show that by elevating intra- and extracellular magnesium ion concentrations, physiological stress protection may be conferred on yeast cells exposed to temperature shock or ethanol toxicity. These and other findings discussed are deemed especially pertinent to high gravity brewing fermentation practices. It is hypothesised that magnesium acts at two levels when stimulating yeast fermentation performance: at the biochemical level to stimulate the activity of glycolytic enzymes; and at the biophysical level to maintain and protect the structural integrity of the yeast plasma membrane.

10.1   Introduction

Fermentative metabolism of brewing yeasts is influenced by the mineral components of wort as well as the available wort sugars. Metal ions which have long been recognised as playing important roles in brewing fermentations are calcium and zinc. The former is involved in depressing wort pH and in governing yeast–yeast flocculation, while deficiency of the latter ion has been shown to be responsible for so-called 'stuck' fermentations. Other metal ions are known to influence yeast growth and metabolism, most notably magnesium. Nevertheless, the roles of magnesium ions in industrial yeast fermentations are not yet fully appreciated[1]. Yeast cells have an absolute requirement for magnesium ions. The essential roles of magnesium in yeast physiology which are pertinent to brewing include: cell division cycle progression[2,3], mitochondrial structure and function[4], respirofermentative metabolism[5], protection against temperature and ethanol-induced stress[6–8] and stimulation of alcohol fermentation rate and yield[9,10]. These latter roles are explained by the essential cofactor functions of magnesium in regulating activities of glycolytic and alcohologenic enzymes.

In contrast to the wide ranging roles of magnesium as discussed above, calcium ions appear to have much less importance for yeast growth and fermentation. For example, calcium is required by yeast cells only at submicromolar levels, as opposed to the millimolar requirements for magnesium. In fact, yeasts actively excrete calcium and maintain extremely low cytosolic levels of this ion. The main roles of calcium are extracellular, for example in governing yeast cell flocculation and as a second messenger in transduction of extracellular signals. Furthermore, calcium ions act antagonistically with magnesium, inhibiting many essential magnesium-dependent biochemical functions[11]. Practical manifestations of such antagonism are evident in industrial fermentations. For example, if the fermentation medium in question, for example brewer's wort, has an unfavourably high Ca:Mg ratio then yeast fermentation performance may be detrimentally affected[9,10,12].

This chapter highlights the roles of metal ions, principally magnesium, in influencing the growth, viability, metabolic vitality and stress tolerance of brewing yeast.

10.2   Materials and methods

10.2.1   *Yeast strains and culture conditions*

Results reported in this paper generally refer to an industrial lager strain of yeast (*Saccharomyces cerevisiae* var. *carlsbergensis*) which was kindly provided by Dr J.A. Hodgson, Scottish Courage Brewing Ltd. Cultures were maintained on malt extract agar slopes at 4°C until required. Pre-culture and fermentation media were either synthetic media described previously[13] or malt wort, kindly supplied by Dr B. Taidi, Scottish Courage Brewing Ltd. Experimental fermentations were conducted in Erlenmyer flasks or simulated tall-tubes (250 ml measuring cylinders). Yeast pitching rates were generally around 10×10^6 cells per ml.

10.2.2   *Yeast growth and viability measurements*

Yeast cell numbers were assessed either by haemocytometer or Coulter counter (Multisizer II). Yeast cell viability was determined either by methylene blue staining or by plate counting of diluted culture aliquots on malt extract agar plates.

10.2.3   *Magnesium preconditioning of yeast*

Elevation of yeast cell magnesium content was achieved by a preconditioning protocol previously described by Walker and Smith[14].

10.2.4   *Imposition of stresses on yeast*

Yeast cells were subjected to temperature- and ethanol-induced stress as described by Walker[6].

10.2.5   *Analytical methods*

Ethanol was determined in culture supernatants by gas chromatography (Hewlett-Packhard 5710A), wort sugars by HPLC (Bio-Rad) and magnesium in cells and

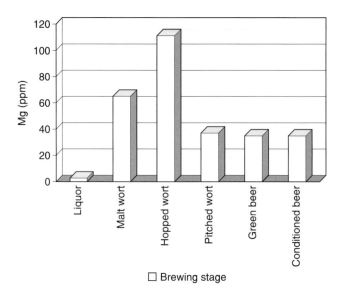

Fig. 10.1 Gains and losses of Mg ions during brewing. Samples were taken from a 300 litre micro-brewery and analysed for total Mg content. The original wort gravity prior to fermentation was 1033 (8°Plato).

media by atomic absorption spectrophotometry (Perkin-Elmer 1100B). Access to a local microbrewery enabled sampling and magnesium analysis of brewing liquids from liquor to beer (see Fig. 10.1).

10.3   Results and discussion

10.3.1   *Magnesium ion availability for brewing yeast*

Brewing yeasts have a very high demand for magnesium ions during fermentation[14] which cannot be met by other metal ions. The supply of magnesium in wort must be sufficient to meet this demand to ensure efficient fermentation performance. This may not always be the case. For example, bioavailable magnesium (that is, free or uncomplexed Mg) may be limiting due to binding, chelation and adsorption by wort constituents and by interactions and biochemical antagonism with other metal ions[11,16]. Figure 10.1 demonstrates the gains and losses of magnesium ions during the brewing process. An important point to make here is that the total magnesium levels measured in wort just prior to pitching may not represent magnesium which is freely available to the yeast. Walker *et al.*[9] have shown that simply by elevating levels of magnesium by wort supplementations, fermentation performance may be enhanced. This additionally ensures that Mg:Ca ratios are maintained at appropriately high levels to minimise the antagonistic effects of calcium on Mg-dependent biochemical processes. An alternative to media magnesium supplementations is the use of Mg-preconditioned yeast.

10.3.2 *Mg-preconditioned brewing yeast*

Walker and Smith[14] and this Volume, Chapter 11, have described an approach to augment yeast cell magnesium levels which may partially counteract any limitations in wort magnesium bioavailability. This involves 'preconditioning' simply by pre-propagating yeast in elevated levels of magnesium (say around 50 mM) prior to the use of washed cells in subsequent fermentations. Figure 10.2 shows that, on an indiviual cellular productivity basis, Mg-preconditioned yeasts are more efficient alcohol producing cell factories compared with unpreconditioned cells. An additional benefit of preconditioning is that Mg-enriched cells appear more tolerant to environmental stress.

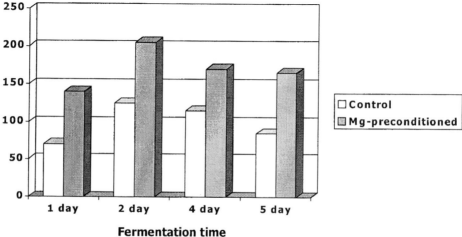

Fig. 10.2 Ethanol productivity of Mg-preconditioned yeast. The yeast cells were preconditioned by propagating in weak wort (OG1020) supplemented with 100 mM Mg and washed in deionised water prior to pitching.

10.3.3 *Magnesium and brewing yeast stress*

Yeast cells subjected to chemical and physical insults lose magnesium ions. Figure 10.3 shows this for ethanol-stressed brewer's yeast, and similar results have been found following temperature shocking of cells[6,7]. By circumventing this loss through magnesium supplementations of the media or by Mg-preconditioning of the cells, yeasts become more stress tolerant. These approaches may be beneficial for high gravity wort fermentations due to the deleterious stresses imposed by both osmotic pressure and ethanol toxicity. It is likely that additional magnesium ions, either in the growth medium or intracellularly, exert their protective effects by charge stabilization of membrane lipids[17]. The yeast plasma membrane therefore becomes more robust and able to maintain its structural integrity in the face of environmental stress caused, for example, by temperature shock or ethanol toxicity.

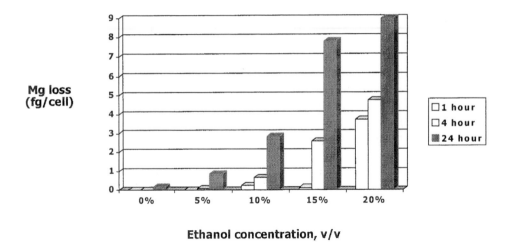

Fig. 10.3   Loss of Mg ions in ethanol stressed brewing yeast. The cells were grown in malt broth, harvested, washed and resuspended in deionised water prior to the addition of ethanol at the concentrations and times indicated. Mg was measured in culture supernatants and expressed on a per cell basis.

10.4   Conclusions

This study has revealed novel, but relatively staightforward, physiological approaches to improving the fermentation performance of brewing yeast based on increasing magnesium ion bioavailability. One approach is to supplement wort with magnesium, which will ensure that high yeast demand for this cation is not compromised by metal ion deficiency, and that inhibitory levels of calcium are minimised. The other approach is to enrich cells with magnesium by preconditioning which, in addition to stimulating fermentative metabolism, will provide yeasts which are more stress tolerant to cope with the rigours of modern brewing practice.

Acknowledgements

The work of the Abertay Yeast Research Group, in particular that of Garry Smith, is gratefully acknowledged. We also thank Scottish Courage Brewing Ltd for supply of materials and Ken Duncan of the Freelance and Firkin, Dundee, for microbrewing facilities.

References

(1)   Walker, G.M. (1984) The roles of magnesium in biotechnology. *Critical Reviews in Biotechnology* **14**, 311–354.
(2)   Duffus, J.H. and Walker, G.M. (1980) Magnesium ions and the control of the cell cycle in yeast. *Journal of Cell Science* **42**, 329–356.
(3)   Walker, G.M. (1986) Magnesium and cell cycle control: an update. *Magnesium* **5**, 9–23.
(4)   Walker, G.M., Birch-Andersen, A., Hamburger, K. and Kramhøft, B. (1982) Magnesium-induced

mitochondrial polymorphism and changes in respiratory metabolism in the fission yeast, *Schizosaccharomyces pombe*. *Carlsberg Research Communications* **47**, 205–214.

(5) Walker, G.M., Maynard, A.I. and Johns, C.G.W. (1990) The importance of magnesium ions in yeast biotechnology. In *Fermentation Technologies: Industrial Applications*, Lu, P.L. (ed.), Elsevier Applied Science, London, pp. 233–240.

(6) Walker, G.M. (1998) Magnesium as a stress protectant for industrial strains of *Saccharomyces cerevisiae*. *Journal of the American Society of Brewing Chemists* **56**, 109–113.

(7) Birch, R.M. and Walker, G.M. (1999) Environmental stress responses in industrial yeasts. In *Proceedings of the Fifth Aviemore Conference on Malting, Brewing and Distilling*, Campbell, I. (ed.), Institute of Brewing, London, pp. 195–199.

(8) Birch, R.M. and Walker, G.M. (2000). Influence of magnesium ions on heat shock and ethanol stress responses of *Saccharomyces cerevisiae*. *Enzyme and Microbial Technology*, in press.

(9) Walker, G.M., Birch, R.M., Chandrasena, G. and Maynard, A.I. (1996) Magnesium, calcium, and fermentative metabolism in industrial yeasts. *Journal of the American Society of Brewing Chemists* **54**, 13–18.

(10) Rees, E.M.R. and Stewart, G.G. (1997) The effects of increased magnesium and calcium concentrations on yeast fermentation performance in high gravity worts. *Journal of the Institute of Brewing* **103**, 287–291.

(11) Walker, G.M. (2000). Biotechnological implications of the interactions between magnesium and calcium. *Magnesium Research*, in press.

(12) Rees, E.M.R. and Stewart, G.G. (1999) Effects of magnesium, calcium and wort oxygenation on the fermentative performance of ale and lager strains fermenting normal and high gravity worts. *Journal of the Institute of Brewing* **105**, 211–217.

(13) Birch, R.M. (1997) PhD thesis, University of Abertay Dundee.

(14) Walker, G.M. and Smith, G.D. (1999) Metal ion preconditioning of brewer's yeast. In *Proceedings of the Fifth Aviemore Conference on Malting, Brewing and Distilling*, Campbell, I. (ed.), Institute of Brewing, London, pp. 311–315.

(15) Walker, G.M. and Maynard, A.I. (1997) Accumulation of magnesium ions during fermentative metabolism in *Saccharomyces cerevisiae*. *Journal of Industrial Microbiology and Biotechnology* **18**, 1–3.

(16) Chandrasena, G., Walker, G.M. and Staines, H.J. (1997) Use of response surfaces to investigate metal ion interactions in yeast fermenations. *Journal of the American Society of Brewing Chemists* **55**, 24–29. 17.

(17) Bara, M., Guiet-Bara, A. and Durlach, J. (1988) Analysis of magnesium membraneous effects: binding and screening. *Magnesium Research* **1**, 29–33.

11 Fermentation Performance of Mg-preconditioned Brewing Yeast

GARRY SMITH and GRAEME WALKER

Abstract   The influence of metal ions in relation to modern day brewing practices has been widely discussed of late, and in particular it has been shown that elevating levels of Mg^{2+} ions in fermentation media stimulates ethanol production and improves stress resistance in industrial strains of *Saccharomyces cerevisiae*. 'Preconditioning' of brewing yeasts at the propagation stage by elevating levels of intracellular Mg is shown here to enhance subsequent fermentation performance of yeast.

11.1   Introduction

With the advent of high gravity brewing, ethanol yields per fermentation have increased, but this may be at the expense of yeast viability and vitality. Brewing yeast physiology may be detrimentally affected by a number of stresses, including osmotic shock, ethanol toxicity, pH/temperature fluctuations and CO_2/hydrostatic pressure[1]. It would therefore be advantageous for modern brewing practice if the metabolic activity and stress tolerance of yeast could be enhanced during the fermentation of high gravity worts.

The roles played by a number of key metal ions in brewing yeast nutrition have been widely discussed[2–7]. These ions are: calcium (which is important in yeast flocculation), magnesium (which is important in enzyme activity and enhancing ethanol output)[4,5,7] and zinc (which is important in activating alcohol dehydrogenase). Their roles are therefore very important for efficient and successful fermentations. In order to redress the problems of reduced yeast viability and vitality during high gravity fermentations, Walker and Smith[8] highlighted the benefits of elevating the bioavailability of magnesium ions to the yeast prior to fermentation.

This chapter describes an approach to counteracting possible limitations in wort magnesium by preconditioning yeast with elevated levels of magnesium ions. Results indicate enhanced fermentation performance of brewing yeasts subject to such treatment.

11.2   Materials and methods

11.2.1   *Yeast strain maintenance*

An industrial lager brewing strain (kindly supplied by Scottish Courage Brewing Ltd) of *Saccharomyces cerevisiae* var. *carlsbergensis* was employed in this study. Cultures were maintained on malt extract agar slopes at 4°C until required.

11.2.2   *Magnesium preconditioning in different media*

Cells were propagated in synthetic medium (EMM3)[9], malt broth (Oxoid) or dilute brewing wort (5°Plato) for 24 h with shaking at room temperature, prior to inoculation into each of the three media with elevated levels of Mg (up to 100 mM). After 24 h and 48 h incubation, cells were thoroughly washed (thrice) with deionised water (18 M conductivity) to remove extraneous magnesium.

11.2.3   *Mg-preconditioned yeast in fermentation*

Cells were precultured in 5°Plato (1020°OG) malt wort for 48 h under aerobic conditions (200 rpm, Schott baffled flasks) at room temperature prior to inoculation into 5°Plato wort with elevated Mg (up to 100 mM). Prior to pitching the cells were washed in sterile distilled water (to remove any loosely bound Mg^{2+} ions from the cell surface). Such preconditioned cells were used as inocula (15×10^6 cells/ml) for high gravity (18°Plato, 1072°OG) experimental fermentations.

11.2.4   *Analytical methods*

Yeast growth was assessed by haemocytometer and dry weight, ethanol by gas chromatography (Hewlett Packard 5710A) and magnesium in yeast cells (following nitric acid hydrolysis of washed cell pellets) by atomic absorption spectrophotometry (Perkin-Elmer 1100B).

11.3   Results and discussion

11.3.1   *Mg-preconditioning*

The effect that propagating cells in various media supplemented with magnesium has on yeast cellular Mg content is shown in Table 11.1. The percentage increase in the amounts of cellular magnesium in preconditioned cells compared with the control cells is indicated. These results indicate that magnesium uptake has been enhanced by yeast cells following preconditioning. However, cell magnesium content is not directly proportional to ionic availability under conditions of Mg sufficiency. In non-limiting magnesium conditions, which would prevail when cells are preconditioned, yeasts accumulate relatively constant amounts of magnesium[10,11]. Nevertheless, significant

Table 11.1   Cell Mg (fg/cell) in preconditioned brewing yeast.[a]

Medium	Control cells (no extra Mg)		Preconditioned cells (50 mM Mg)		Preconditioned cells (100 mM Mg)	
	24 h	48 h	24 h	48 h	24 h	48 h
Synthetic medium	100	40	166 (66)	106 (165)	236 (136)	188 (370)
Malt broth	70	74	190 (171)	163 (120)	325 (364)	156 (111)
Brewing wort	131	137	174 (33)	207 (51)	204 (56)	255 (86)

[a] Data shown in parentheses represent % increase in cell Mg of control.

increases in Mg accumulation by brewing yeast cells can be achieved by preconditioning.

11.3.2 Fermentation performance of preconditioned cells

Cells propagated in unsupplemented wort show an uptake in Mg after 24 h of fermentation, indicating that they still require extra magnesium for growth (Fig. 11.1). Yeasts then gradually uptake and release Mg during fermentation[10], a phenomenon also noted by Mochaba et al.[3] By increasing the levels of intracellular Mg prior to fermentation the need for cells to take up the ion during the first 24 h of fermentation may be reduced[3]. Cells which are Mg-preconditioned may accumulate and metabolise sugars predominantly for ethanol production, rather than for growth and cell division. However, a basal level of magnesium will be required to govern progress of the yeast cell division cycle[12].

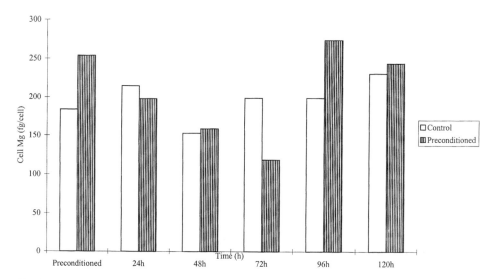

Fig. 11.1 Yeast cell Mg content during fermentation. The cells were propagated in weak wort (5°Plato) without (control) or with (preconditioned) Mg supplementation. Mg levels in the propagation wort were 1.3 mM and 100 mM, respectively.

Preconditioned cells may have enhanced vitality and may exhibit significant Mg release upon pitching into fresh wort. Such release is associated with a change in membrane permeability followed by subsequent magnesium uptake which also occurs concomitantly with sugar uptake[13]. Mochaba et al.[3] suggested that increased wort Mg, observed immediately upon pitching, may represent a good indicator of brewing yeast vitality. Our findings concur with the concept that yeast cell magnesium ion homeostasis and fermentation performance are inextricably linked.

11.4 Conclusions

The results show that yeast cells propagated in the presence of elevated magnesium levels are more fermentatively active and yield more ethanol than unpreconditioned yeasts. These results may be beneficial in brewing yeast management, especially with regard to high gravity fermentation practices.

Acknowledgements

We thank the ESF for funding and Scottish Courage Brewing Ltd for yeast strain and wort supply.

References

(1)   Walker, G.M. (1998) Magnesium as a stress protectant for industrial strains of *Saccharomyces cerevisiae*. *Journal of the American Society of Brewing Chemists* **56**, 109–113.

(2)   Jones, R.P. and Greenfield, P.F. (1984) A review of yeast ionic nutrition. Part 1: Growth and fermentation requirements. *Process Biochemistry* April, 48–58.

(3)   Mochaba, F., O'Connor-Cox, E.S.C. and Axcell, B.C. (1996) Metal ion concentration and release by a brewing yeast: characterisation and implications. *Journal of the American Society of Brewing Chemists* **54**, 155–163.

(4)   Rees, E.M.R. and Stewart, G.G. (1997) The effects of increased magnesium and calcium concentrations on yeast fermentation performance in high gravity worts. *Journal of the Institute of Brewing* **103**, 287–291.

(5)   Rees, E.M.R. and Stewart, G.G. (1999) Effects of magnesium, calcium and wort oxygenation on the fermentative performance of ale and lager strains fermenting normal and high gravity worts. *Journal of the Institute of Brewing* **105**, 211–217.

(6)   Walker, G.M. and Maynard, A.I. (1996) Magnesium limited growth of *Saccharomyces cerevisiae*. *Enzyme and Microbial Technology* **18**, 455–459.

(7)   Walker, G.M., Birch, R.M., Chandrasena, G. and Maynard, A.I. (1996) Magnesium, calcium and fermentative metabolism in industrial yeasts. *Journal of the American Society of Brewing Chemists* **54**, 13–18.

(8)   Walker, G.M. and Smith, G.D. (1999) Metal ion preconditioning of brewer's yeast. In *Proceedings of The Fifth Aviemore Conference on Malting, Brewing and Distilling*, Campbell, I. (ed.), The Institute of Brewing, London, pp. 311–315.

(9)   Birch, R.M. (1997) Ph.D. thesis, University of Abertay Dundee.

(10) Walker, G.M. and Maynard, A.I. (1997) Accumulation of magnesium ions during fermentative metabolism in *Saccharomyces cerevisiae*. *Journal of Industrial Microbiology and Biotechnology* **18**, 1–3.

(11) Saltukoglu, A. and Slaughter, J.C. (1983) The effect of magnesium and calcium on yeast growth. *Journal of the Institute of Brewing* **89**, 81–83.

(12) Walker, G.M. and Duffus, J.H. (1980) Magnesium ions and the control of the cell cycle in yeast. *Journal of Cell Science* **42**, 329–356.

(13) Stephanopoulus, D. and Lewis, M.J. (1968) Release of phosphate by fermenting brewer's yeast. *Journal of the Institute of Brewing* **74**, 378–383.

12   Stress During Fermentation

PETER PIPER, G. DUNNE and BILL LANCASHIRE

Abstract   There are very good reasons for believing that the performance of yeast in fermentations could be enhanced by minimising the stress exposure of the yeast cells. A large number of genes in the yeast are stress genes, genes characteristically responsive to specific stresses (e.g. high ethanol, osmotic stress, oxidative stress, etc.). By monitoring the expression of these genes in brewing fermentations we can get a very good idea of which specific stresses the yeast cells are experiencing. This in turn should indicate which alterations to the fermentation might minimise stress exposure of the yeast, thereby enhancing yeast performance.

This work has focused mostly on one ale strain and one lager strain. It indicates that the best characterised stress genes of laboratory yeasts while still present in these brewing yeasts have very different reponsiveness to stress. Probably the lager strain has at some stage acquired part of the genetic background of a psychrophilic yeast. Analysis of stress gene mRNAs at different timepoints of fermentations conducted with these ale and lager strains indicates that the major response of the cells is to the ethanol stress, not to osmostress.

13 The Oxidative Stress Response of Ale and Lager Yeast Strains

VERONIQUE MARTIN, DAVID QUAIN and KATHERINE SMART

Abstract Fermentation performance and beer quality are influenced by yeast physiological state. During serial repitching, yeast cells might be exposed to stress conditions as a result of fermentation or storage handling. Free radicals (e.g. superoxide anion) and reactive oxygen species (e.g. hydrogen peroxide and hydroxy radical) are generated from yeast aerobic metabolism and are known components of fermenting wort. These compounds damage cellular macromolecules such as lipids, proteins and DNA, representing a significant stress to brewing yeast during fermentation. Although, the primary oxidant defences are provided by enzymes such as superoxide dismutases (SOD1 and SOD2) and catalases (CTT1 and CTA1), other non-enzymatic antioxidants may also provide protection (e.g. glutathione). It is suggested that resistance to oxidants may influence yeast physiological condition, resistance to other stresses and eventually fermentation performance.

Production ale and lager strains grown in YPD or wort to achieve populations exhibiting exponential or stationary phase have been exposed to oxidants including menadione (a generator of superoxide anions) and hydrogen peroxide. Resistance to oxidative stress was observed to be strain dependent and affected by media composition and growth phase. The antioxidant defence levels have been investigated for these strains. The levels of glutathione, superoxide dismutases and catalases expressed are strain dependent and affected by media composition and growth phase. The sensitivity of brewing yeast strains to oxidative stress will vary with yeast handling procedures and, in particular, direct oxygenation of yeast slurries. As the levels of oxidant compounds tend to be greater during high gravity fermentations, strains which exhibit greater resistance to oxidative damage or more effective stress response mechanisms may be more suitable for high gravity brewing.

13.1 Introduction

Oxidative stress comprises the stress associated with cells responding to and protecting themselves from the reactive oxygen species[1]. Reactive oxygen species (superoxide anions [O_2^-], hydrogen peroxide [H_2O_2] and hydroxy [OH^-] radicals) may result in damage to cellular macromolecules such as lipids, proteins and DNA[7,8] and are mainly generated during yeast aerobic metabolism[3,4]. Although brewing fermentations are essentially anaerobic, yeast cells are exposed to oxygen at pitching[5] and during storage[6] representing a potential source of oxidative stress. Cell defences against reactive oxygen species are provided by enzymes (superoxide dismutases and catalases) and other non-enzymatic antioxidants (glutathione, metal ions, vitamins C and E)[1,3,4]. It is suggested that oxidative stress may affect yeast quality and subsequently fermentation performance.

The tolerance of production strains of ale and lager yeast to exogenously generated oxidative stress in the form of hydrogen peroxide and superoxide anions has been investigated. The relationship between the resistance to these oxidants and the cellular antioxidant defence (total catalase and superoxide dismutase activities and cellular glutathione concentration) has been examined for ale and lager strains grown in YPD.

13.2   Materials and methods

13.2.1   *Yeast strains and growth conditions*

Three lager (BB10, BB11 and BB28) and two ale (BB3 and BB18) production strains were obtained from Bass Brewers. Yeast cells were grown aerobically in 250 ml flasks containing 50 ml of YPD (yeast extract 1% w/v, bacteriological peptone 2% w/v, glucose 2% w/v) at 25°C on an orbital shaker at 120 rpm. Yeast populations were harvested in mid-exponential or stationary phase.

13.2.2   *Determination of response to oxidative stress*

Yeast cells were washed three times in sterile deionised water and resuspended in either water, H_2O_2 (0.01, 0.1 and 1% w/v) or menadione (10, 50, 100, 200 mM) to a final cell concentration of 1×10^5 cells per ml and incubated at 25°C on an orbital shaker at 120/min. Cell viability was assessed using plate counts on YPD and expressed as a percentage viability.

13.2.3   *Glutathione concentration*

Yeast cells were harvested by centrifugation and washed three times in cold (4°C) sterile deionised water. Cells were resuspended in 500 µl metaphosphoric acid 5% (w/v). Cell suspensions were repeatedly boiled and immersed in liquid nitrogen (six times) to achieve cell lysis. The supernatant was retained for analysis. Glutathione concentration was determined using the assay kit from Calbiochem (La Jolla, CA, USA) and expressed as nmol per cell concentration.

13.2.4   *Protein extraction for enzymatic assays by glass bead cell lysis method*

Yeast cells were harvested by centrifugation and washed three times in cold (4°C) sterile deionised water. Cells were resuspended in 100 µl fresh buffer (Tris pH 7.5 50 mM, EDTA 0.5 mM, Triton X100 0.1% (v/v) and the following protease inhibitors suspended in DMSO 100%: TPCK (tosylphenylalanine chloromethyl ketone) 0.25 mM, TLCK (tosyllysine chloromethyl ketone) 0.25 mM or in absolute methanol PMSF (phenylmethylsulphonyl fluoride) 0.5 mM). To this suspension 100 µl of acid washed glass beads (0.5 mm) were added. The samples were vortexed for 3 min, centrifuged at $2100 \times \mathbf{g}$ for 10 min and the resulting supernatant retained for analysis. The total protein concentration from each extract was determined using a Bio-Rad protein assay kit (Bio-Rad Laboratories, GmbH, München, Germany).

13.2.5   Catalase activity

Protein extracts were obtained following the glass bead lysis method. Cellular catalase activity was determined according to the methods of Aebi[7] and Izawa et al.[8] The catalase activity was expressed as units per g protein.

13.2.6   Superoxide dismutase activity

Protein extracts were obtained following the glass bead lysis method. Cellular superoxide dismutase activity was determined using the assay kit from Calbiochem (La Jolla, CA, USA) and expressed as SOD-525 units per mg protein.

13.3   Results and discussion

13.3.1   Oxidative stress tolerance for YPD grown brewing yeast cells

For haploid strains of *Saccharomyces cerevisiae* the response of YPD grown exponential and stationary phase cells to H_2O_2 and the superoxide anion generator menadione has been established[4], with stationary phase cells exhibiting greater resistance than exponential phase cells. YPD grown brewing yeast cells also exhibited an increased resistance for stationary phase populations and a reduced resistance for exponential phase cells following exposure to exogenous H_2O_2[9].

The resistance of YPD grown brewing yeast cells to menadione was also growth phase dependent (Figs 13.1 to 13.4). Indeed, brewing yeast cells grown to stationary phase demonstrated relatively high viabilities, whereas total cell death was observed for exponential phase cells after only 2 h exposure to 100 mM menadione (Figs 13.1 and 13.2). Viability for all strains decreased with the duration of exposure to either H_2O_2 or menadione, and the rate of cell death was dependent on the concentration of oxidant. For brewing yeast grown in YPD, H_2O_2 tolerance was reported to be strain dependent with the ale strains being more sensitive than the lager strains for both exponential (Fig. 13.3) and stationary phase populations[9] (Fig. 13.4). The effect of superoxide anion stress on viability was also strain dependent for both exponential (Fig. 13.1) and stationary (Fig. 13.2) phase populations; ale strains were more sensitive to this oxidant than lager strains, but the difference was not as pronounced as that observed following exposure to H_2O_2.

13.3.2   Antioxidant defences

The primary defences against oxidative stress are provided by enzymes such as superoxide dismutases (*SOD1* and *SOD2*) and catalases (*CTT1* and *CTA1*). However, other non-enzymatic antioxidants (such as glutathione, metal ions, vitamins C and E) may also protect the cell[1,3,4].

13.3.2.1   Superoxide dismutase activity expressed by stationary phase ale and lager strains.   The yeast *S. cerevisiae* contains two superoxide dismutase genes, *SOD1* encoding for the cytoplasmic Cu/Zn SOD and *SOD2* encoding Mn SOD located in

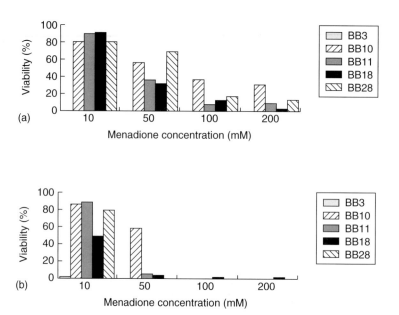

Fig. 13.1 Effect of oxidative stress on the viability of YPD grown lager and ale exponential phase yeast populations following exposure to menadione (10, 50, 100 and 200 mM) for (a) 1 hour and (b) 2 hours. Viability values represent the mean of three replicate samples.

the mitochondrial matrix[3]. Superoxide dismutase enzymes play an important role in the disproportionation of the superoxide anion into H_2O_2 and dioxygen, and are mainly expressed during stationary phase of growth. The *SOD1* gene may be regulated by copper availability[10], whereas *SOD2* is repressed by cAMP and induced by exogenous oxidants[11].

Superoxide dismutase activity for ale (BB1) and lager (BB11) brewing yeast strains has been reported previously following aerobic growth on semi-defined wort at 18°C and 12°C, respectively[12], though superoxide dismutase activity of brewing yeast strains has not been reported previously following aerobic growth on YPD. The superoxide dismutase activity of stationary phase populations of the three lager and two ale strains of brewing yeast grown on YPD was observed to be strain dependent (Table 13.1). However, superoxide dimutase levels did not correlate with the resistance of brewing strains to either of the oxidants used. The reasons for this are not known.

13.3.2.2 *Catalase activity expressed by stationary phase ale and lager strains.* S. *cerevisiae* cells contain two catalase genes, *CTA1* and *CTT1*, which encode the peroxisomal catalase A and the cytosolic catalase T enzymes, respectively. Catalases are primarily expressed during stationary phase of growth and catalyse the removal of H_2O_2 (by hydrolysis to form water and oxygen) during this phase[13]. Both genes are induced by oxygen; however, catalase A gene expression may be induced by certain fatty acids and by growth on non-fermentable carbohydrate, but it may be strongly

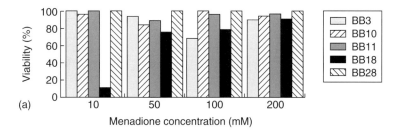

(a)

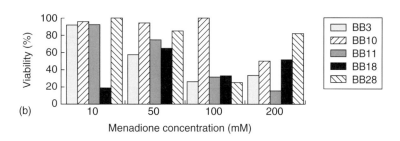

(b)

Fig. 13.2   Effect of oxidative stress on the viability of YPD grown lager and ale stationary phase yeast populations following exposure to menadione (10, 50, 100 and 200 mM) for (a) 1 hour and (b) 2 hours. Viability values represent the mean of three replicate samples.

Table 13.1   Antioxidant defences for stationary phase cells grown aerobically in YPD (the data are expressed as the mean $\pm$ standard deviation and represents the mean of at least six replicates).

Strains	Glutathione concentration	Catalase activity	SOD activity
BB3	57.9 $\pm$ 16.5	9.0 $\pm$ 6.4	11.1 $\pm$ 1.1
BB10	68.6 $\pm$ 27.0	41.9 $\pm$ 6.8	13.4 $\pm$ 8.3
BB11	78.6 $\pm$ 40.5	27.0 $\pm$ 5.3	5.9 $\pm$ 2.7
BB18	43.3 $\pm$ 6.5	56.2 $\pm$ 13.9	4.1 $\pm$ 1.1
BB28	73.9 $\pm$ 4.7	29.5 $\pm$ 2.8	4.3 $\pm$ 3.2

repressed by glucose. The catalase T gene is negatively regulated by cAMP and has been shown to be induced by stresses including starvation, osmotic and oxidative stress[14]. Catalase activity is therefore strain, media composition and stress dependent[1].

Catalase activity for ale (BB1) and lager (BB11) brewing yeast strains have been previously examined following aerobic growth on semi-defined wort at 18°C and 12°C, respectively[12]. The total catalase activity of stationary phase populations of the three lager and two ale strains of brewing yeast was reported to be strain dependent in

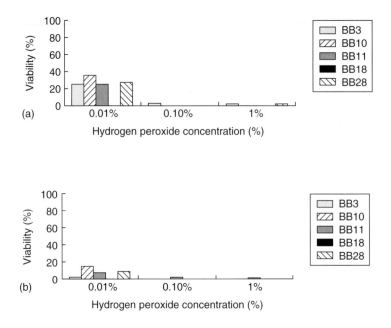

Fig. 13.3 Effect of oxidative stress on the viability of YPD grown lager and ale exponential phase yeast populations following exposure to hydrogen peroxide (0.01, 0.1, 1% v/v) for (a) 1 hour and (b) 2 hours. Viability values represent the mean of three replicate samples.

both YPD and wort, though the influence of growth media on the levels of this enzymatic antioxidant defence did not appear to be universal or consistent[9]. However, the catalase levels observed (Table 13.1) reflect the relative resistance of each strain to 1% (v/v) exogenous H_2O_2 (Figs 13.3 and 13.4), but not at the lower concentrations where the reduction in viability is less pronounced. No relationship was observed between the tolerance to menadione and the levels of catalase activity.

13.3.2.3 *Levels of glutathione expressed by stationary phase ale and lager strains.*

Apart from enzymatic defences, *S. cerevisiae* cells produce a non-enzymatic antioxidant, glutathione, which reduces H_2O_2 to water and oxygen[8]. Glutathione is present in two forms in yeast cells, the reduced antioxidant form (GSH) and the oxidised form (GSSG), and is important for many biological processes[4]. The oxidised form (GSSG) is recycled through a reaction catalysed by the enzyme glutathione reductase[15].

The level of total cellular glutathione in YPD grown stationary phase populations of lager and ale brewing yeast strains has been previously reported; it was observed that the glutathione concentration exhibited in both YPD and wort grown cells were strain dependent[9]. Indeed, YPD grown ale strain cells exhibited the lowest levels of glutathione (Table 13.1). Typically the levels of glutathione in the cell reflect the level of endogenously generated oxidative stress imposed on the cells or impaired gluta-

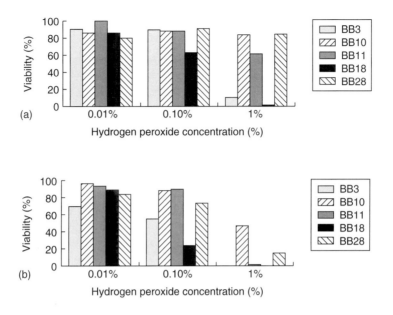

Fig. 13.4 Effect of oxidative stress on the viability of YPD grown lager and ale stationary phase yeast populations following exposure to hydrogen peroxide (0.01, 0.1, 1% v/v) for (a) 1 hour and (b) 2 hours. Viability values represent the mean of three replicate samples.

thione reductase activity resulting in the necessity to generate higher levels of the antioxidant.

13.4 Conclusions

The tolerance of brewing yeast strains to exogenous H_2O_2 and menadione stress was dependent on strain and the phase of growth exhibited by the cell population. Lager strains appeared to be more resistant than ale strains, though the reason for this is not known.

The levels of superoxide dismutase activity reflect the cellular defence against superoxide anion stress and the activity was observed in this study to be strain dependent. However, superoxide dismutase activity did not correlate with either menadione or H_2O_2 stress tolerance.

Cellular catalase activity and glutathione content indicate the level of defence against H_2O_2, and were found to be strain dependent. Although the catalase levels appeared to be directly related to the resistance of the strains to oxidative stress, the glutathione content of the cells was inversely related to the activity of this enzyme; glutathione may therefore compensate for reduced catalase activity and vice versa. Indeed, a recent study suggested that glutathione and catalase represent redundant H_2O_2 detoxification systems[16].

The relationship between antioxidant potential and fermentation performance remains the subject of further investigation.

Acknowledgements

Veronique Martin is supported by the Henry Mitchell Scholarship. The authors would like to thank the Directors of Bass Brewers for kind permission to publish this work.

References

(1)　Santoro, N. and Thiele, D.J. (1997) In *Yeast Stress Responses*, Hohmann, S. and Mager, W.H. (eds), Springer-Verlag, Heidelberg, Chap. 6.

(2)　Fridovitch, I. (1978) The biology of free radicals. *Science* **201**, 875–879.

(3)　Gralla, E.B. and Kosman, D.J. (1992) Molecular genetics of superoxide dismutases in yeasts and related fungi. *Adv. Genet.* **30**, 251–319.

(4)　Jamieson, D.J. (1998) Oxidative stress responses of the yeast *Saccharomyces cerevisiae*. *Yeast* **14**, 1511–1527.

(5)　Boulton, C.A. and Quain, D.E. (1987) *Proceedings of the European Brewery Convention Congress*, Madrid **21**, 401–408.

(6)　Boulton, C.A. (1991) Yeast management and the control of brewery fermentations. *Brewers Guardian* April, 25–29.

(7)　Aebi, H. (1984) Catalase *in vitro*. *Methods Enzymol* **105**, 121–126.

(8)　Izawa, S., Inoue, Y. and Kimura, A. (1995) Oxidative stress response in yeast: effect of glutathione on adaptation to hydrogen peroxide stress in *Saccharomyces cerevisiae*. *FEBS Letters* **368**, 73–76.

(9)　Martin, V., Quain, D.E. and Smart, K.A. (2000) The oxidative stress response of ale and lager yeast strains. *Proceedings of the European Brewery Convention Congress*, in press.

(10)　Lapinskas, P.J., Ruis, H. and Culotta, V.C. (1993) Regulation of *Saccharomyces cerevisiae* catalase gene expression by copper. *Current Genet* **24**, 388–393.

(11)　Flattery-O'Brien, J.A., Grant, C.M. and Dawes, I.W. (1997) Stationary-phase regulation of the *Saccharomyces cerevisiae SOD2* gene is dependent on additive effects of HAP2/3/4/5- and STRE-binding elements. *Molec. Microbiol.* **23**, 303–312.

(12)　Clarkson, S.P., Large, P.J., Boulton, C.A. and Bamforth, C.W. (1991) Synthesis of superoxide dismutases, catalases and other enzymes and oxygen and superoxide toxicity during changes in oxygen concentration in cultures of brewing yeast. *Yeast* **7**, 91–103.

(13)　Izawa, S., Inoue, Y. and Kimura, A. (1996) Importance of catalase in the adaptive response to hydrogen peroxide: analysis of acatalasaemic *Saccharomyces cerevisiae*. *Biochem. J.* **320**, 61–67.

(14)　Ruis, H. and Hamilton, B. (1992) *Molecular Biology of Free Radical Scavenging Systems*, Cold Spring Harbor Laboratory Press, NY, pp. 153–172.

(15)　Collinson, L.P. and Dawes, I.W. (1995) Inducibility of the response of yeast cells to peroxide stress. *Gene* **156**, 123–127.

(16)　Grant, C.M., Perrone, G. and Dawes, I.W. (1998) Glutathione and catalase provide overlapping defenses for protection against hydrogen peroxide in the yeast *Saccharomyces cerevisiae*. *Biochem. Biophys. Res. Commun.* **253**, 893–898.

14   The Death of the Yeast Cell

KATHERINE SMART

Abstract 'Ageing' is a term which describes the progressive deleterious change in the physiological state of a production yeast during the brewing process. For the brewing industry the term 'yeast ageing' is non-specific and refers to any population of cells which has been exposed to a range of life-experiences. Indeed, there are several means by which yeast may become and consequently be described as aged. Thus brewing yeast ageing refers to four distinct physiological states: stationary phase; stored yeast populations; serially repitched yeast populations; and individual cells exhibiting the 'true' aged phenotype.

Brewing yeast ageing is the predetermined progressive transition of an individual cell from youth to old age that finally culminates in the death of the cell. Yeast ageing is a function of the number of divisions undertaken by an individual cell, and may be measured by enumerating the number of bud scars on the cell surface. Individual brewing yeast cells are mortal and exhibit a finite lifespan, determined by genetic and environmental factors.

An aged yeast cell exhibits morphology and physiology distinct from those of younger cells, although the impact of this modification on cellular function and fermentation performance has yet to be elucidated. Cellular markers for senescence, include a linear increase in size with age, the accumulation of surface wrinkles, an increase in bud scar number, and terminal cell lysis. Since a rough cell surface topography favours cell to cell adhesion during the onset of flocculation, and it would be expected that discrete older wrinkled cells would be more flocculant than discrete younger smoother cells.

Towards the end of the yeast lifespan mother cells are slower to divide and the daughters produced may be retained by their parent. Although the reasons for this are unclear, is has been postulated that this could be due to ineffective separation of bud and birth scar, rather than a reduction in the rate of cytokinesis. The impact of a propensity to retain daughter cells during the latter stages of lifespan on fermentation performance is unclear, but it is postulated that such cells may form the nuclei for floc formation much in the same way as chain-forming brewing yeast strains behave. Where the pitch consists primarily of aged cells, an extended lag phase in the fermenter may result due to the slow progression through the cell cycle.

14.1   Introduction

The metabolic and physiological events leading to cell death have predominantly been investigated in higher organisms. Although it is accepted that repeated exposure to stress during yeast handling and fermentation often results in the occurrence of deteriorated and dying cells, the processes involved in the initiation of death metabolism have not been fully elucidated. One reason for this is that individual cell deaths regularly occur and are often masked by the replication of surviving cells or represent a minimal and in brewing terms an acceptable reduction in population viability.

Beer quality is strongly influenced by the biochemical performance of the yeast during fermentation. A dying cell exhibiting aberrant metabolism would be unlikely to perform well during fermentation, and cell death which culminates in autolysis may subsequently affect beer clarity and flavour.

14.2   The causes of brewing yeast cell death

The progressive nature of the cellular deterioration leading to death has not been clearly defined, however, it is known that the following forms of physiological state[1] occur: (i) replicating healthy cell; (ii) non-replicating cell; (iii) reversible deterioration; (iv) irreversible deterioration – cell is dying; (v) non-viable cell; and (vi) autolysis (Fig. 14.1). Such deterioration may result from necrosis (either as a result of external stress or DNA damage leading effectively to lethal or suicidal genotypes) or senescence (predetermined finite replicative lifespan) (Table 14.1).

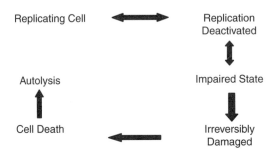

Fig. 14.1   Schematic representing the transition from live to death states.

Table 14.1   Summary of stresses that may lead to brewing yeast cell death (adapted from Walker (1998)[2]).

Physical stress	Chemical stress	Biological stress
Cold shock	Yeast metabolites	Lethal DNA damage
Osmotic shock	Starvation	'Suicide mutations'
Shear	Free radicals	Senescence
Pressure	pH	
Gaseous	Metal ions	

14.3   The yeast cell cycle

A replication activated cell is able to progress through the cell cycle. Yeast cultures including those employed in brewery fermentations comprise cells at varying stages of the cell cycle[3]. The normal yeast cell cycle involves the progression through a series of landmark events that enable the cell to complete a round of cellular reproduction known as division. From the brewing perspective this process is critical for the formation of sufficient yeast biomass to generate the required product as well as sufficient crop in the cone for subsequent harvesting and reuse. One of the most important landmark events during the cell cycle is known as 'start'. 'Start' resides in the G_1 phase of the cycle (see Fig. 14.2) and is an event that commits the yeast cell to division. It is during the G_1 phase that the decision to enter 'start' and thus progress through a new round of cell division is made[4]. Failure to progress through 'start', due to the absence

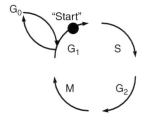

Fig. 14.2   The yeast cell cycle (adapted from Smart, 1999[5]).

of nutrients, results in cell division arrest and entry into the cell cycle stationary phase, which is also known as G_0. At this stage the cell is considered to exhibit replication deactivation, though this may readily be reversed provided that the cells do not progress to an irreversibly damaged state (Fig. 14.1).

14.4   Identifying compromised states

Identifying cells of compromised state is essential in the determination of pitching rates. In addition, yeast condition reflects the efficiency of yeast handling and may assist the brewer to predict the potential performance of the slurry.

During the yeast life cycle there are several compromised physiological states which may culminate in loss of fermentative potential either temporarily or permanently. The capacity of a culture to exhibit replication activation is most readily characterised by the occurrence of budding cells. Indeed during exponential growth the proportion of cells which have progressed through 'start' and those which have not is roughly equal and may be represented by the occurrence of 50% budded and 50% unbudded cells.

Thus replication deactivation may be identified by measuring the budding index in liquid culture. Permanent deactivation is more readily assessed by monitoring divisional capacity on solid media using either micromanipulation[6], slide counts or plate counts. The transition from replication deactivation to impaired physiological state (Fig. 14.1) may be identified by assessing the activity or vitality of the cell. Various markers of activity that have been used to indicate vitality are reviewed in the earlier chapters of this volume. The accumulation of irreversible damage leading to a loss of cell viability may be identified using a variety of techniques, including brightfield and fluorescence stains (see Chapter 1 for a review).

14.5   Brewing yeast ageing

'Ageing' is a term which describes the progressive deleterious change in the physiological state of a production yeast during the brewing process[5]. For the brewing industry the term 'yeast ageing' is non-specific and refers to any population of cells which has been exposed to a range of life-experiences. Indeed, there are several means by which yeast may become, and consequently be described as, aged. Thus brewing yeast ageing refers to four distinct physiological states: stationary phase; stored yeast populations; serially repitched yeast populations; and individual cells exhibiting the 'true' aged phenotype[5].

14.6   Stored yeast populations

For most breweries yeast storage is an essential intermediate stage between cropping and pitching. Ideally, the storage regime utilised should allow the physiological state of the cropped yeast to be maintained[7], although deterioration can occur, particularly following prolonged exposure to various stresses including aerobiosis, anaerobiosis, nutrient starvation, ethanol, low pH and temperature abuse[8–10].

At the point of cropping the yeast slurry exhibits stationary phase due to both an absence of oxygen and severely limited availability of nutrients. Thus the majority of the cropped population will appear unbudded, and will reside in the G_0 phase of the cell cycle. This is fortuitous in some respects because, in general, stationary phase yeast cells are more resistant to physiological stress than their exponential equivalents. Nevertheless, the stresses imposed during storage can be extreme[9] and subsequent fermentation performance may well be adversely affected.

Therefore yeast biomass may have a variable physiological status prior to pitching, and although the impact of this on fermentation performance is not fully understood recent observations have shown that the 'physiological history' of the cell may affect its subsequent efficiency[10,11] and product quality[12]. Because yeast which has been stored for extended periods exhibits both the stationary phase and stressed phenotype, it is often perceived to be aged. Perhaps a more accurate description for such a population would be deteriorated or physiologically impaired[5].

14.7   Serial repitching

Good yeast handling is an essential prerequisite for brewery fermentations. During the cycle of serial repitchings the yeast may be exposed to various physiological stresses[13–15], resulting in slurries of variable condition[13] and subsequent inconsistent fermentation performance[14]. Little is known about the effect of extended serial repitching on subsequent fermentation performance and cropped yeast condition. However, it is known that deterioration does result from this form of 'ageing'. The resultant modifications are progressive and the physiology of the crop affects the condition of subsequent slurries[14]. It would appear, though, that extended serial repitching does constitute a form of 'repetitive stress injury' in which the yeast population is repeatedly transferred from the cell cycle to stationary phase to stress conditions and back to the cell cycle[5]. Interestingly, extended exposure to this cycle of events results in the progressive deterioration of the yeast biomass.

14.8   Individual cell ageing – a definition

For the brewer, then, an aged yeast usually refers to a crop which has been stored for extended periods or a pitch which has previously been used for several successive

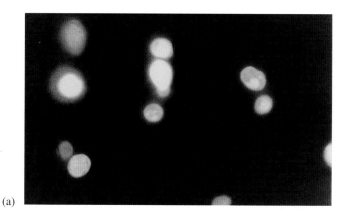

(a)

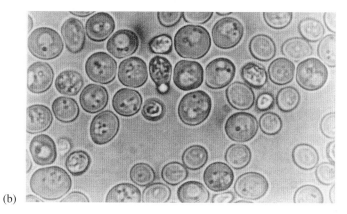

(b)

Plate 1 Organisms from 17 day cider yeast culture stained with oxonol examined by (a) fluorescent microscopy and (b) phase contrast microscopy.

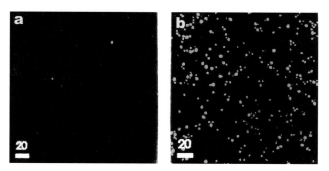

Plate 2 Effect of heat treatment on oxonol permeability in cider yeast: (a) live and (b) heat killed (>65°C, 15 min).

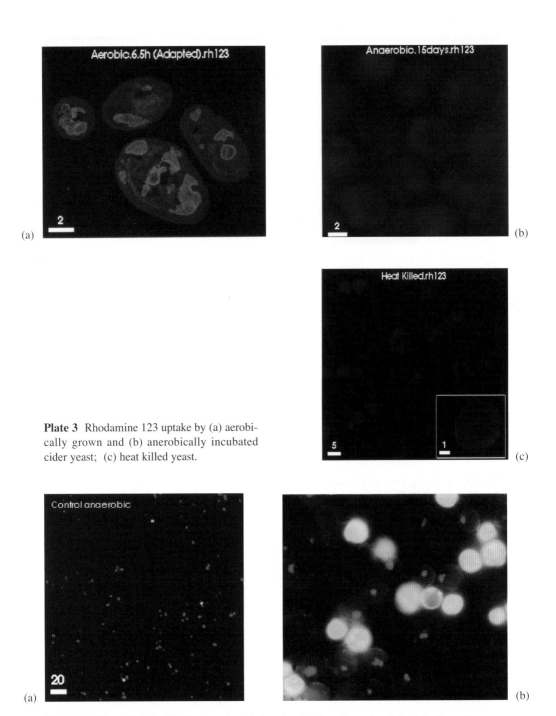

Plate 3 Rhodamine 123 uptake by (a) aerobically grown and (b) anerobically incubated cider yeast; (c) heat killed yeast.

Plate 4 (a) Propidium iodide fluorescein staining showing live (green) and dead (red) yeasts, compared with the results using (b) the Fungolight kit from Molecular Probes.

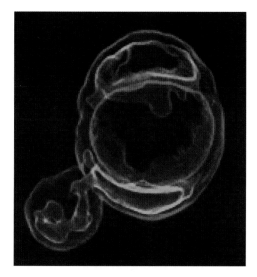

Plate 5 (top) Yellow-green fluorescent intracellular staining and red-orange inter-vacuolar structures .

Plate 6 (middle) Typical methylene blue staining of yeast.

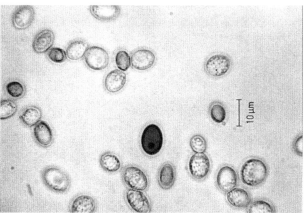

Plate 7 (bottom) Reproducibility of PCR reactions for detection of lactic acid bacteria. One batch of chemically extracted DNA and two batches of enzymatically isolated DNA from pure cultures of lactic acid bacteria were used to test the reproducibility of PCR reactions with the primer set 907r and LB-95. Each set of four reactions was repeated in duplicate. The primer set 907r and LB-95 was then used to detect the presence of unidentified lactic acid bacteria in samples of DNA mechanically extracted from a contaminated yeast slurry. The control organisms were *Lactobacillus paracasei paracasei* (isolate 1), *L. paracasei paracasei* (isolate 2), *L. paracasei paracasei* (isolate 3), and *Pediococcus pentosauceus* (isolate 4). The reactions (all with the primer set 907r and LB-95) were as follows.

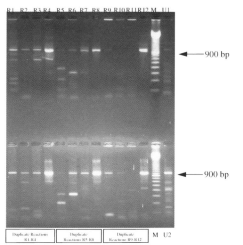

R1 Chemically extracted DNA from a pure culture of isolate 1

R2 Chemically extracted DNA from a pure culture of isolate 2

R3 Chemically extracted DNA from a pure culture of isolate 3

R4 Chemically extracted DNA from a pure culture of isolate 4

R5 Enzymatically extracted DNA (batch 1) from a pure culture of isolate 1

R6 Enzymatically extracted DNA (batch 1) from a pure culture of isolate 2

R7 Enzymatically extracted DNA (batch 1) from a pure culture of isolate 3

R8 Enzymatically extracted DNA (batch 1) from a pure culture of isolate 4

R9 Enzymatically extracted DNA (batch 2) from a pure culture of isolate 1

R10 Enzymatically extracted DNA (batch 2) from a pure culture of isolate 2

R11 Enzymatically extracted DNA (batch 2) from a pure culture of isolate 3

R12 Enzymatically extracted DNA (batch 2) from a pure culture of isolate 4

U1 Chemically extracted DNA from an artificially contaminated yeast slurry

U2 Duplicate reaction of U1

M   Molecular weight ladder (800 pb band brightest)

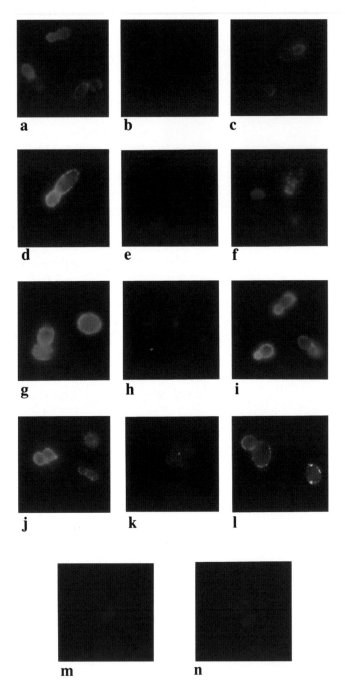

Plate 8 Yeast phenotype identification using cell surface Flo lp characterised by immunofluorescent staining.

fermentations. In fact this form of population chronological ageing does not truly relate to the individual age of the yeast cell.

Brewing yeast cells possess a finite replicative lifespan[6,16,17]. The number of replications or divisions an individual cell can complete is dependent on both genetic and environmental factors[16,18,19]. The maximum division capacity of a cell is termed the Hayflick limit[20]. Once cells have reached their Hayflick limit, they are incapable of further replication and enter a physiological state termed senescence, which leads to cell death (Fig. 14.3).

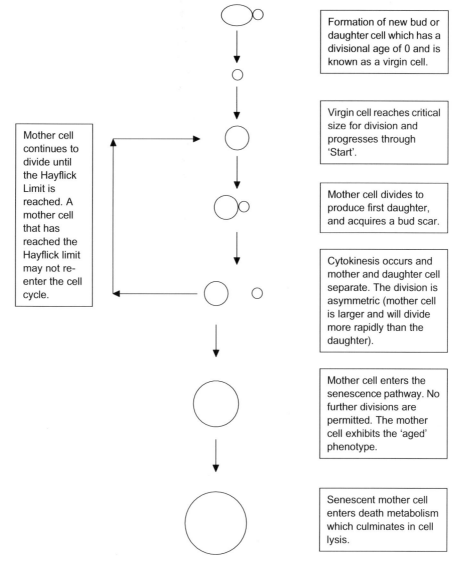

Fig. 14.3 Progression through lifespan in yeast is a function of the number of divisions undertaken (adapted from Smart, 1999[5]).

14.9   Strain longevity

Since 1950 the phenomenon of individual cell ageing in yeast has been examined by a few hardy microbiologists[21]. To determine the Hayflick limit, a cohort of no less than 60 cells must be individually monitored to establish the number of divisions undertaken. This process requires the use of a microscope with tetrad stage and micromanipulator[6]. Individual cells are monitored throughout their replicative lifespans, from newly emerged bud to senescent mother cell. By recording the number of divisions accomplished by each cell, age specific mortality profiles for each strain are obtained.

Using this method, yeast cellular senescence has been demonstrated in haploid[22–24], diploid[25] and polyploid[6] strains. A new yeast daughter cell or bud becomes a mother cell that can divide to produce 10 to 33 daughters of its own, depending on the strain[19].

14.10   A state of senescence

For brewing yeast, ageing results in modifications in appearance which are irreversible. Several morphological and physiological changes associated with the ageing process have been described. These include (1) an increase in bud scar number and therefore chitin deposition[6,26]; (2) an increase in cell size[6]; (3) accumulation of surface wrinkles[6,16]; (4) eventual cell lysis[27]; and (5) increased retention of daughter cells[27].

An extensive increase in cell size occurs throughout the lifespan[6], with newly produced daughter cells exhibiting a small mean cell size of around 150 μm^3 compared with the mean volume of 950 μm^3 typically demonstrated by senescing cells. Indeed, cell size was observed to increase following each division (see Fig. 14.4), and the rate of enlargement was linear throughout the lifespan. Interestingly, the increase in size of the mother cell did not result in a corresponding increase in size for each successive bud produced[27], except in very rare instances when the newly formed bud could equal the size of its ageing mother.

Penultimate and ultimate daughters are invariably retained by their mother cells for longer than those produced at earlier stages in the lifespan[6] and a propensity to form chains is concomitant with a reduced lifespan[27]. The impact of a propensity to retain daughter cells during the latter stages of lifespan is unclear, but it is postulated that such cells may form the nuclei for floc formation much in the same way as chain-forming brewing yeast strains behave. Throughout most of the lifespan, the time taken to progress through the cell cycle is uniform. However, during the last 20% of the lifespan, the division time dramatically increases[6], probably due to inefficient and slow progression through G_1 to 'start'. Thus if the pitch consisted primarily of aged cells, an extended lag phase in the fermenter would result due to the slow progression through the cell cycle. If the pitch consisted primarily of newly budded virgin cells, the time taken to reach the critical size required for the first division would result in a slight delay in the onset of growth. Therefore, mother cells that are young to middle aged represent the most active portion of the pitch population in terms of rapid growth and yeast biomass production. In addition, once 80 to 90% of the lifespan has been completed, the cells exhibit a granular appearance accompanied by wrinkling

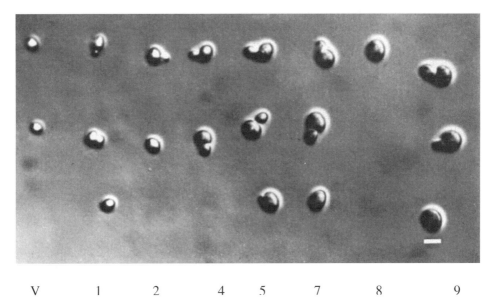

V   1   2   4   5   7   8   9

Fig. 14.4 Relationship between divisional age and cell size (reproduced with permission from the *Journal of the American Society of Brewing Chemists*).

(Fig. 14.5) at the cell surface[6]. This manifestation of cell wall modifications is supported by the observation that the cell surface characteristics change with replicative age. Cell surface charge, which typically inversely correlates to flocculation potential[28], decreases with increase in divisional age (Table 14.2). Since a rough cell surface topography and reduced cell surface charge favour cell to cell adhesion during the onset of flocculation[10,11,29], it would be expected that discrete older wrinkled cells would be more flocculant than discrete younger smoother cells.

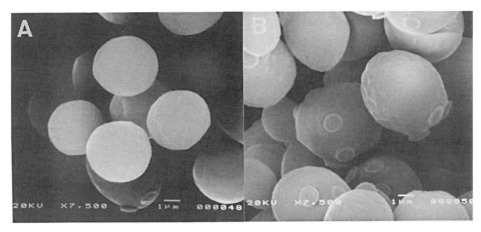

Fig. 14.5 Cell surface topography of A, virgin and B, five-to-six-division old cells (reproduced with permission from the *Journal of the American Society of Brewing Chemists*).

Table 14.2 The impact of divisional cell age on cell surface charge.

Cell population	Alcian blue retention (μg/mg YDW)
Exponential	55
Stationary	23
Virgin	38
Young mothers (2 to 3)	46
Older mothers (4 to 8)	29

14.11 Individual yeast cell ageing and serial repitching

At the end of fermentation and sometimes also during yeast storage, cells are allowed to settle, resulting in a sediment at the bottom of the vessel. This settling process is known to be zonal in nature with larger flocs settling faster than smaller flocs, and larger discrete cells settling more rapidly than their smaller companions. Thus in conical vessels it is possible that older cells may sediment more rapidly than their younger counterparts due either to their propensity to retain daughters and form flocs or their increased cell size. Serially cropping and re-pitching enriched populations of older cells could inadvertently result in the selection of aged subpopulations, with implications for subsequent fermentation performance. Therefore an important practical consideration for the brewer is which portion of the yeast 'cone' to retain and how may sequential fermentations to pitch with the same yeast culture. In a normal population there should always be 50% newly budded daughters exhibiting only their birth scar, 25% one division old yeast exhibiting their birth scar and one bud scar, 12.5% exhibiting their birth scar and two bud scars, 6.25% exhibiting their birth scar and three bud scars and so on. Thus if enrichment in the cone does occur, enumerating the number of bud scars on individuals from a given portion within the cone should demonstrate a tendency towards older cells in the lower regions of the sediment. Deans et al.[26] suggested that older cells were indeed more abundant in lower rather than higher regions of the cone, supporting the original hypothesis of Barker and Smart[6].

14.12 Conclusions

The transition from youth to senescence is not chronological in yeast. No matter how active the yeast is in terms of fermentation, lifespan is related to the number of cell divisions it has undertaken and not to any other parameter. Even so ageing appears to be marked by a progressive impairment in cellular mechanisms, resulting in irreversible changes in physiology and morphology. It can only be assumed that this is accompanied by a loss of dynamic range in physiological function, and consequently a reduced capacity to adapt to stress and ferment[5].

References

(1) Jones, R.P. (1987) Measures of yeast death and deactivation and their meaning: Parts I and II. *Process Biochemistry* **22–23**, 1118–128.

(2) Walker, G.M. (1998) *Yeast Physiology and Biotechnology*, Wiley, Chichester.

(3) Futcher, B. (1996) Cyclins and the wiring of the yeast cell cycle. *Yeast* **12**, 1635–1646.

(4) Wheals, A.E. (1987) *The Yeasts*, Vol. 1, Rose, A.H. and Harrison, J.S. (eds), Academic Press, London, pp. 283–390.

(5) Smart, K.A. (1999) Ageing in brewing yeast. *Brewers Guardian* **128** (2), 19–24.

(6) Barker, M.G. and Smart, K.A. (1996) Morphological changes associated with the cellular ageing of a brewing yeast strain. *Journal of the American Society of Brewing Chemists* **54** (2), 121–126.

(7) Sall, C.J., Seipp, J.F. and Pringle, A.T. (1988) Changes in brewers yeast during storage and the effect of these changes on subsequent fermentation performance. *Journal of the American Society of Brewing Chemists* **46**, 23–25.

(8) Knusden, F.B. (1985) *Journal of the American Society of Brewing Chemists* **43** (2), 91–95.

(9) Monch, D., Kruger, E. and Stahl, U. (1995) *Monastsschr. Brauwiss.* **48** (9/10), 288–299.

(10) Smart, K.A., Boulton, C.A., Hinchliffe, E. and Molzahn, S. (1995) Effect of physiological stress on the surface properties of brewing yeasts. *Journal of the American Society of Brewing Chemists* **53** (1), 33–38.

(11) Rhymes, M.R. and Smart, K.A. (1996) Effect of starvation on the flocculation of ale and lager brewing yeasts. *Journal of the American Society of Brewing Chemists* **54**(1), 50–56.

(12) Morimura, H., Hino, T., Kida, K. and Maemura, H. (1998) Storage of pitching yeast for the production of whisky. *Journal if the Institute of Brewing* **104**(4), 191–244.

(13) Boulton, C. A. (1991) Yeast management and the control of brewery fermentations. *Brewers Guardian* **120**(3), 25–29.

(14) Smart, K.A. and Whisker, S.W. (1996) Effect of serial repitching on the fermentation properties and condition of brewing yeast. *Journal of the American Society of Brewing Chemists* **54**, 41–44.

(15) O'Connor- Cox, E. (1998) Improving yeast handling in the brewery. Part 2: Yeast collection. *Brewers Guardian* **126**(12), 26–34.

(16) Rodgers, D.L., Kennedy, A.I., Hodgson, J.A. and Smart, K.A. (1999) Lager brewing yeast ageing and stress tolerance. In *Proceedings of the European Brewery Convention Congress* **27**, 671–678.

(17) Powell, C.D., Quain, D.E. and Smart, K.A. (2000) The impact of media composition and petite mutation on the longevity of a polyploid brewing yeast strain. *Letters in Applied Microbiology*, in press.

(18) Barker, M.G., Brimage, L. and Smart, K.A. (1999) Effect of Cu,Zn superoxide dismutase disruption mutation on ageing in *Saccharomyces cerevisiae*. *FEMS Letters,* **177**, 199–204.

(19) Powell, C.D., Van Zandycke, S.M., Quain, D.E. and Smart, K.A. (2000) Replicative ageing and senescence: a review. *Microbiology*, in press.

(20) Hayflick, L. (1965) The limited *in vitro* lifetime of human diploid cell strains. *Experimental Cell Research* **37**, 614–636.

(21) Barton, A.A. (1950) *Journal of General Microbiology* **65**, 84–86.

(22) Austriaco Jr., N.R. (1996) Review – To bud until death: the genetics of ageing in the yeast *Saccahromyces cerevisiae*. *Yeast* **12**(7), 623-630.

(23) Egilmez, N.K. and Jazwinski, S.M. (1989) Evidence for the involvement of a cytoplasmic factor in the aging of the yeast *Saccharomyces cerevisiae*. *Journal of Bacteriology* **171**, 37–42.

(24) Muller, I. (1971) Experiments on ageing in single cells of *Saccharomyces cerevisiae*. *Arch Mikrobiol.* **77**, 20–25.

(25) Mortimer, R.K. and Johnston, J.R. (1959) Life span of individual yeast cells. *Nature* **183**, 1751–1752.

(26) Deans, K., Pinder, A., Cately, B.J. and Hodgson, J.A. (1997) Effects of cone cropping and serial repitch on the distribution of cell age in brewery yeast. *Proceedings of the European Brewing Congress* **26**, 469–476.

(27) Powell, C.D., Quain, D.E. and Smart, K.A., unpublished.

(28) Smart, K.A. (2000) *Flocculation and adhesion*. Monographs in Yeast Physiology, European Brewing Convention, in press.

(29) Day, A.W., Poon, N.H. and Stewart, G.G. (1975) Fungal fimbrae. III: The effect on flocculation in *Saccharomyces*. *Canadian. Journal of Microbiology* **21**, 558–564.

15 The Relationship Between Brewing Yeast Lifespan, Genetic Variation and Fermentation Performance

CHRIS POWELL, DAVID QUAIN and KATHERINE SMART

Abstract   Limited studies of brewing yeast ageing have been reported, but the relationship between lifespan and fermentation performance has not been investigated. Lifespan analysis was performed using seven production strains selected for their brewing classification (lager/ale) and morphological characteristics (chain-forming/discrete). In addition, a petite mutant and a flocculent variant were analysed for longevity. It was observed that each strain displayed a unique ageing profile in terms of its mean and maximum lifespan, indicating that, in brewing yeast, longevity is a strain specific phenomenon. Mitochondrial function was observed to influence mean and maximum[10%] lifespan potential.

15.1   Introduction

Ageing may be defined as the gradual increase in deleterious intracellular modifications which occur continually throughout the lifespan leading to senescence and finally death[1–3]. This phenomenon is accompanied by a decrease in metabolic activity and vitality and by characteristic physiological changes[3–5]. Brewing yeast lifespan has been investigated previously only for lager yeast strains[6–8]. It has been postulated that lifespan is determined by an organism's genes and influenced by environmental factors[3,5]; however, the relationship between longevity in polyploid yeast and mitochondrial function has not previously been investigated. Due to stresses imposed on yeast during the brewing process, the frequency of genetic drift and mutations is high[9,10]. Pitching wort with respiratory deficient yeast can result in poor fermentation performance[9,10]. Here we report on the relationship between lifespan in ale and lager production strains and the impact of petite mutation on brewing yeast ageing and replicative senescence.

15.2   Materials and methods

15.2.1   Yeast strains

Seven proprietary strains were provided by Bass Brewers, Burton (Table 15.1). A petite mutant of the lager strain BB11, designated BB11p and a flocculent variant of BB11, denoted BB56, were also obtained (Table 15.1).

15.2.2   Media and growth conditions

Each strain was maintained and propagated on YPD media (2% w/v bacteriological peptone, 1% w/v yeast extract, 2% w/v glucose), 1.2% agar was added when solid media were required. Media were sterilised immediately following preparation by

Table 15.1   Longevity of *S. cerevisiae* strains. Lifespan is expressed as the mean and maximum[10%] (upper decile population) replicative lifespan. The standard deviations (S.D.) from these values are shown. All strains were provided by Bass Brewers.

Strain	Ploidy	Type	Cells manipulated	Mean lifespan ± S.D.	Maximum[10%] lifespan ± S.D.
BB1	Polyploid	Ale strain	73	14.3 ± 5.3	22.9 ± 1.8
BB3	Polyploid	Ale strain	70	10.3 ± 4.7	19.6 ± 3.5
BB10	Polyploid	Lager strain	68	9.7 ± 5.7	22.4 ± 4.0
BB11	Polyploid	Lager strain	215	33.0 ± 10.7	50.6 ± 4.1
BB18	Polyploid	Ale strain	74	21.7 ± 7.5	31.3 ± 1.6
BB28	Polyploid	Lager strain	71	19.9 ± 9.2	35.0 ± 2.1
BB62	Polyploid	Lager strain	68	18.2 ± 8.4	34.3 ± 4.6
BB11p	Polyploid	Petite of BB11	71	19.4 ± 12.0	38.7 ± 3.8
BB56	Polyploid	Flocculent variant	70	32.2 ± 9.3	46.9 ± 3.0

autoclaving at 121°C and 15 psi for 15 min. All media components were supplied by Oxoid.

15.2.3   Determination of cell age

The Hayflick limit of each strain was established using micromanipulation. The method described by Barker and Smart[6] was applied for cell age determination.

15.2.4   Data analysis

The data obtained from micromanipulation were expressed as the mean and maximum[10%] (mean Hayflick limit of the upper decile population) of the lifespan. Significance was determined using the heteroscedastic Student *t*-test, and was deemed to have been demonstrated when the value did not exceed 0.05 (5% confidence level).

15.3   Results and discussion

15.3.1   Ageing and senescence in brewing yeast is a strain specific phenomenon

The Hayflick limit in *Saccharomyces cerevisiae* has been postulated to be species and strain specific[1,11,12], although no extensive study has been published. In order to examine this hypothesis, the lifespans of seven brewing yeast strains were examined, comprising four lager and three ale strains. Five of these strains exhibited discrete and the remaining two partial chain forming morphology (Table 15.1).

It was observed that each yeast strain displayed a unique ageing profile in terms of its mean and maximum lifespan (Fig. 15.1, Table 15.1). Each strain investigated exhibited a constant rate of acceleration in mortality (Fig. 15.1), indicating a typical Gompertz distribution[1,13]. Analysis of the mean and maximum[10%] Hayflick limits for each strain indicated that no two strains exhibited identical longevities (Table 15.1). Although the individual lifespans of a cohort of cells represent a normal distribution around the mean, the range of this distribution is strain specific such that two strains

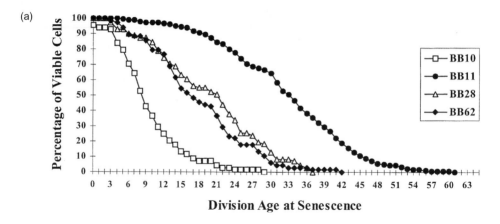

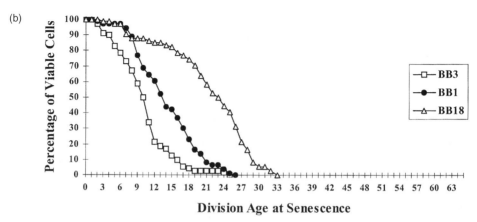

Fig. 15.1   Mortality profiles of (a) four production lager yeast strains and (b) three production ale yeast strains. Lifespans of cohorts of cells were monitored from emergent bud through to the senescent state. Strain specific mortality profiles, including the range of lifespan over which senescence occurs, are illustrated.

which exhibit similar mean lifespans may differ with respect to their maximum[10%] lifespan (Table 15.1). The greatest replicative lifespan displayed was observed for the polyploid ale yeast BB28, which demonstrated a maximum[10%] lifespan of 35.0 ± 2.1 divisions and a mean lifespan of 21.7 ± 7.5 divisions, and the shortest longevity was exhibited by the ale yeast BB3, which only replicated a maximum[10%] of 19.6 ± 3.5 times prior to senescence with a mean lifespan of 10.3 ± 4.7 divisions (Table 15.1).

15.3.2   *The impact of petite mutation on yeast longevity*

It has been suggested that mitochondrial function is essential for longevity[14]. Analysis of haploid petite mutants have shown a 40%[15] and a 25%[16] reduction in lifespan; however, there has been no previous reported analysis of brewing yeast petite mutant lifespan.

In order to study the impact of loss of respiration capacity on yeast longevity, the lifespan of a petite mutant, BB11p, derived from BB11 was investigated on YPD. Lifespan data for BB11 and its petite may be found in Table 15.1. When compared with BB11, the petite strain displayed a reduced lifespan ($P < 0.05$). BB11p exhibited a mean lifespan of 19.4 $\pm$ 8.4, which corresponds to a 40% reduction in lifespan when compared with the original BB11 strain (32.4 $\pm$ 10.4). The flocculent variant BB56 was not observed to differ from the original in either mean or maximum[10%] lifespan ($P < 0.05$) (Table 15.1, Fig. 15.2). It is postulated that the formation of respiratory deficient mutants during yeast handling may impact on yeast longevity. The altered ageing profile observed in petite mutants may be a causative factor behind the poor fermentation performance observed when wort is pitched with respiratory deficient yeast.

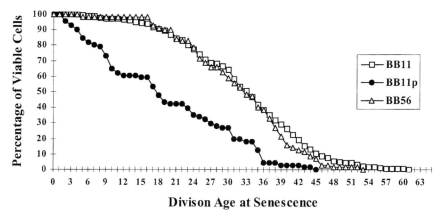

Fig. 15.2 Mortality profiles of BB11, a flocculent variant BB56 and a respiratory deficient petite mutant BB11p. Lifespans of cohorts of cells were monitored from emergent bud through to the senescent state.

15.4 Conclusions

Seven proprietary polyploid brewing strains, exhibiting discrete or chain forming cell morphology were analysed for longevity on YPD media. Each strain displayed a unique pattern of ageing in terms of its mean and maximum[10%] lifespan. In addition, differences in mortality profiles indicated that the point at which the maximal rate of mortality is initiated and the age band during which cell death peaks within the population is also strain dependent. These observations have not previously been reported, and suggest that ageing and senescence in brewing yeast are strain specific phenomena.

The lager strain BB11 was selected as the subject for further analysis of yeast ageing and senescence. A flocculent variant of BB11 was observed to exhibit a similar lifespan to the original strain, suggesting that in this strain variation in flocculation genes does not result in reduced longevity assurance mechanisms. A respiratory deficient mutant of BB11 was also examined for longevity to determine the impact of mitochondria on senescence. It was observed that the petite mutant exhibited a reduced lifespan when compared with the wild type, supporting previous observations[15]. It is suggested that mitochondrial integrity is essential to assure brewing yeast lifespan potential.

Acknowledgements

The authors are grateful to the Directors of Bass Brewers for permission to publish. Chris Powell is a Rainbow scholar funded by the BBSRC and Bass Brewers.

References

(1) Austriaco, N.R. (1996) Review – to bud until death: the genetics of ageing in the yeast, *Saccharomyces. Yeast* **12**, 623–630.
(2) Hayflick, L. (1965) The limited *in vitro* lifespan of human diploid cell strains. *Exp. Cell Res.* **37**, 614–636.
(3) Jazwinski, S.M. (1990) Ageing and senescence of the budding yeast *Saccharomyces cerevisiae. Molec. Microbiol.* **4**, 337–343.
(4) Jazwinski, S.M. (1993) The genetics of ageing in the yeast *Saccharomyces cerevisiae. Genetica* **91**, 35–51.
(5) Sinclair, D.A., Mills, K.A. and Guarente, L. (1998) Ageing in *Saccharomyces cerevisiae. Annu. Rev. Microbiol.* **52**, 533–560.
(6) Barker, M.G. and Smart, K.A. (1996) Morphological changes associated with the cellular ageing of a brewing yeast strain. *J. Amer. Soc. Brew. Chem.* **54**, 121–126.
(7) Deans, K., Pinder, A., Catley, B.J. and Hodgson, J.A. (1997) Effects of cone cropping and serial re-pitch on the distribution of age distribution of cell ages in brewery yeast. *Proceedings of the European Brewing Congress* 469–476.
(8) Rodgers, D.L., Kennedy, A., Hodgson, J.A. and Smart, K.A. (2000) Ageing and stress tolerance in lager brewing yeast. *Proceedings of the European Brewing Congress*, in press.
(9) Morrison, K.B. and Suggett, A. (1983) Yeast handling, petite mutants and lager flavour. *J. Inst. Brew.* **89**, 141–142.
(10) Stewart, G. (1996) Yeast performance and management. *The Brewer* **83**, 211–215.
(11) Egilmez, N.K. and Jazwinski, S.M. (1989) Evidence for the involvement of a cytoplasmic factor in the aging of the yeast *Saccharomyces cerevisiae. J. Bacteriol.* **171**, 37–42.
(12) Muller, I. (1971) Experiments on ageing in single cells of *Saccharomyces cerevisiae. Arch. Mikrobiol.* **77**, 20–25.
(13) Pohley, H.J. (1987) A formal mortality analysis for populations of unicellular organisms. *Mech. Ageing Dev.* **38**, 231–243.
(14) Grant, C.M., MacIver, F.H. and Dawes, I.W. (1997) Mitochondrial function is required for resistance to oxidative stress in the yeast *Saccharomyces cerevisiae. FEBS Lett.* **410**, 219–222.
(15) Berger, K.B. and Yaffe, M.P. (1998) Prohibition family members interact genetically with mitochondrial inheritance components in *Saccharomyces cerevisiae. Molec. Cell Biol.* **18**, 4043–4052.

16 The Influence of Media Composition on Replicative Lifespan in Lager Brewing Yeast

DAWN RODGERS, ALAN KENNEDY, PAT THURSTON,
JEFF HODGSON and KATHERINE SMART

Abstract   Beer quality is strongly influenced by the biochemical performance of the yeast during fermentation. The rate and extent of fermentation and the quality of the final product may be affected by both extrinsic and intrinsic factors. Lager brewing yeast cells exhibit a finite replicative lifespan which is strain dependent with characteristic morphological changes associated with age and senescence.

Lifespan in haploid strains of *Saccharomyces cerevisiae* has been demonstrated to be affected by media composition. To elucidate the impact of media components on polyploid yeast lifespan, the effect of fermentable carbon source in defined media on the longevity of four lager brewing yeast strains has been investigated. It is suggested that media composition may moderate brewing yeast lifespan, though the extent of change is strain dependent.

16.1   Introduction

All eukaryotic cells possess a finite replicative lifespan[1]. The number of replications or divisions an individual cell can complete is termed the Hayflick limit[2]. Once cells have reached their Hayflick limit, they are incapable of further replication and enter a physiological state termed senescence, which leads to cell death. A new yeast daughter cell or bud becomes a mother cell that can divide to produce 13 to 30 daughters of its own, depending on the strain[3]. Barker and Smart[4] demonstrated that lager brewing yeast exhibit a finite replicative lifespan, subsequently Powell *et al.*[5] have shown that lager and ale brewing yeast longevities differ and are strain dependent.

For yeast, ageing also results in modifications in appearance which are irreversible. Several morphological and physiological changes associated with the ageing process in one lager brewing yeast have been described[4]. These include: an increase in bud scar number and therefore chitin deposition; an increase in cell size; a granular appearance to aged cells; the accumulation of surface wrinkles; the retention of daughter cells towards the end of lifespan; and eventual cell lysis.

Yeast longevity is a function of both genetic and environmental factors[1]. Environmental factors which have been postulated to affect lifespan include oxidative stress[6], nutrient deficiency and starvation (reviewed by Smart[3]). Barker *et al.*[7] have demonstrated that the carbon source on which yeast strains are grown affects lifespan for haploid strains of *Saccharomyces cerevisiae*. Growth on glycerol or glucose resulted in similar mean and maximum Hayflick limits, but growth on galactose resulted in an extension in both mean and maximum lifespan. Growth on ethanol has been demonstrated to affect yeast lifespan for haploid[7], diploid[8] and polyploid[9] yeast strains, though the modifications observed were strain dependent[9]. Here we demonstrate that lifespan is a strain dependent phenomenon which may be affected by media composition.

16.2   Materials and methods

16.2.1   *Yeast strains*

Four production strains of lager brewing yeast designated SCB1, SCB2, SCB3 and SCB4 were obtained from Scottish Courage Technical Centre, Edinburgh.

16.2.2   *Media and growth conditions*

Each strain was maintained and grown on YP medium (2% w/v bacteriological peptone, 1% w/v yeast extract) with either glucose, maltose or maltotriose as a carbon source at a concentration of 2% w/v. Wort (1060°) was obtained from Scottish Courage Brewers, Berkshire Brewery, Reading. Where required 1.2% agar was added.

16.2.3   *Micromanipulation*

YP plates no more than 5 mm thick were inoculated with a single yeast colony and incubated for 48 h at 25°C. The resulting microcolony was then examined using a Zeiss microscope with a long working distance 40× objective lens, by viewing through the Petri dish and agar. Cells were manipulated using a micromanipulation glass needle. Virgin cells were isolated by the separation of newly formed buds away from midsized mother cells. Careful monitoring of cell cycle progression and subsequent separation of newly generated daughter cells allowed the development from virgin to aged mother cells to be investigated. Plates were incubated at 25°C during the day and at 4°C overnight to decrease growth rate and prevent excessive division. Where necessary, filter paper soaked in sterile deionised water was placed in the lid of each Petri dish to prevent desiccation of the media. More than 50 cells were monitored for each strain.

16.2.3.1   *Data analysis.* The data obtained from the micromanipulation was expressed as the mean and maximum$^{10\%}$ (mean Hayflick limit of the upper decile of the population of the lifespan). In addition to this, the standard deviation for each sample group was calculated. Significance was determined using the heteroscedastic Student *t*-test, and was deemed to be demonstrated when the value did not exceed 0.05 (5% confidence level).

16.3   Results and discussion

16.3.1   *Determination of brewing yeast lifespan on YP with glucose as the sole carbon source*

Yeast replicative lifespan is known to be strain dependent[1,3] and has been postulated to fall within 10 to 30 divisions. The replicative lifespans of five lager brewing yeast strains with different ancestral backgrounds have been reported[4,9]. The mean and maximum$^{10\%}$ lifespan of each strain are presented in Table 16.1.

Table 16.1 Mean and maximum[10%] lifespans for lager brewing yeast strains (S.D. = standard deviation). The strains were grown on YP with glucose (2% w/v) agar and maintained at 25°C during the day and 4°C overnight.

Strain	Total cells manipulated	Mean lifespan	S.D.	Max. lifespan	Max.[10%] lifespan	S.D.[10%]	Reference
SCB1	122	15.51	7.20	32	27.83	1.70	9
SCB2	165	9.54	4.83	29	18.17	3.91	9
SCB3	98	26.43	11.60	55	45.40	5.46	9
SCB4	125	12.36	5.45	28	20.46	3.23	9
L1116	—	17.20	—	29	—	—	3

The longevities obtained (Table 16.1) demonstrate that each of the four yeasts in this investigation exhibits its own unique ageing or mortality profile (Fig. 16.1), supporting the theory that ageing is a strain specific phenomenon. None of the four yeast strains examined was found to have a maximum lifespan greater than those which had previously been published, although SCB3 demonstrated a maximum lifespan equal to that of PSY142, a long-lived haploid yeast strain, examined by Austriaco[1]. The lifespan data for the lager brewing yeast strain LL1116[4], when compared with the data for the four strains presented, exhibited an intermediate lifespan. The mean lifespan demonstrated by SCB2 was very low, with no comparable data having been published. Indeed, this is the lowest lifespan yet reported for a *S. cerevisiae* yeast strain. The mean replicative lifespan data were analysed to determine the relationship between strain and longevity. Statistical analysis using the heteroscedastic Students *t*-test demonstrated that there was a significant difference ($P < 0.05\%$) between the four yeast strains under investigation, indicating that lifespan potential is under genetic control. Although it was noted that the age related changes in morphological characteristics, such as increase in cell size and the occurrence of surface wrinkles during the latter stages of lifespan, were conserved (Table 16.2)[9], the timing of onset and extent of each characteristic were strain dependent (Table 16.2). The mechanism by which these changes occur has yet to be fully elucidated.

16.3.2 *The impact of carbon source on replicative senescence*

The carbohydrate profile of brewers wort comprises fermentable (glucose, fructose, sucrose, maltose and maltotriose) and non-fermentable sugars. Glucose, fructose and sucrose, which are present in relatively low concentrations, are rapidly utilised by brewer's yeast. Maltose, which represents up to 50 to 55% of the total carbohydrate concentration of the wort[10,11] may only be utilised once the glucose is exhausted, because of catabolite repression[12,13]. The relationship between yeast lifespan and maltose or maltotriose has not previously been investigated. Maltotriose represents up to 15 to 20% of the total carbohydrate concentration of the wort[14] and is utilised once the maltose is exhausted. Since these carbon sources represent the most abundant carbohydrate components of brewers wort, the replicative lifespan on maltose and maltotriose was investigated for all four lager strains.

It has been suggested that yeast lifespan is the function of both genetic background and environmental factors[1]. The carbon source on which yeast strains are grown has been demonstrated to affect haploid yeast lifespan[7], though the effect of carbon source on polyploid and industrial strains has not been previously reported. In order to determine the impact of fermentative carbon source available during brewery fermentations on brewing yeast longevity, the replicative lifespan exhibited by the lager strains, SCB1, SCB2, SCB3 and SCB4 were investigated on maltose and maltotriose. It was observed that the effect of maltose and maltotriose utilisation on brewing yeast lifespan was strain specific (Tables 16.3 and 16.4). The mean Hayflick limit of SCB1 on maltose was equivalent to that observed on glucose, SCB2 exhibited

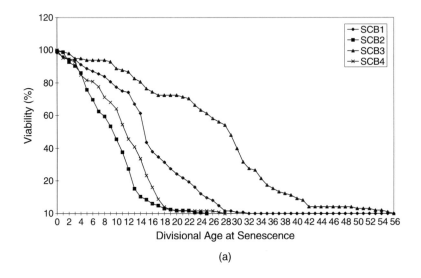

(a)

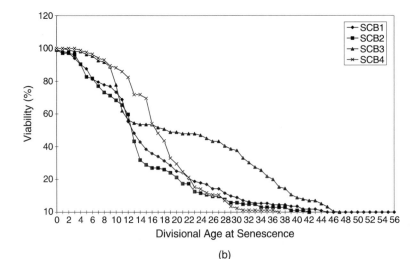

(b)

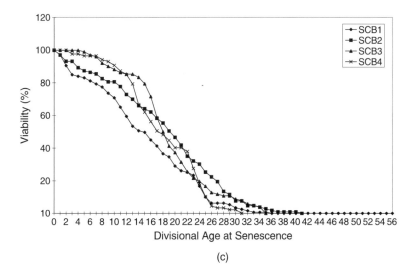

(c)

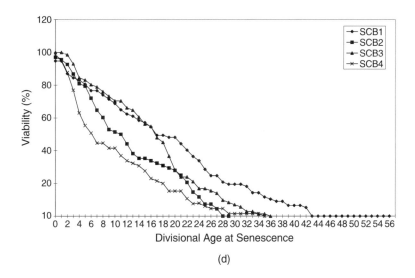

(d)

Fig. 16.1 Mortality profiles for four lager brewing yeast strains grown on YP with (a) glucose, (b) maltose, (c) maltotriose and (d) brewer's wort as sole carbon source.

an increased lifespan and SCB3 a decreased longevity. SCB4 demonstrated increased mean and maximum[10%] lifespans, although the maximum[10%] lifespan of this strain was the shortest observed on maltose. The reasons for this strain dependent effect are not known and merit further investigation.

The maximum lifespans (Table 16.4) determined on maltotriose were all greater than those on glucose except SCB3. This increase was not as pronounced as that observed on maltose. The same pattern was observed when comparing the maximum[10%] lifespan data. Growth on maltotriose resulted in increased mean lifespans

Table 16.2 Morphological characteristics observed during replicative lifespan for the four lager brewing yeast strains on YPD (adapted from Rodgers et al.[9]).

Characteristic	SCB1	SCB2	SCB3	SCB4
Cell size	large, increasing throughout the lifespan	large, increasing throughout the lifespan	very small original size, increases over lifespan	large, rapid observable increase in cell size
Division age where surface wrinkles occur	4	6	15 to 20	6
Initiation of senescent phenotype as a function of lifespan	14%	33%	39%	29%
Propensity for retaining daughter cells	little	slightly greater frequency, chains formed, but, easily broken	little	little
Division time	2 to 2.5 h	2 to 2.5 h	> 3 h	2 to 2.5 h

Table 16.3 Mean and maximum[10%] lifespans for lager brewing yeast strains (S.D. = standard deviation). The strains were grown on YP with maltose (2% w/v) agar and maintained at 25°C during the day and 4°C overnight (adapted from Rodgers et al.[9]).

Strain	Total cells manipulated	Mean lifespan	S.D.	Max. lifespan	Max.[10%] lifespan	S.D.[10%]
SCB1	112	15.76	9.73	45	36.64	5.57
SCB2	104	14.27	8.61	42	33.30	5.17
SCB3	107	22.36	13.07	46	44.75	1.39
SCB4	86	17.75	6.38	37	29.89	2.85

Table 16.4 Mean and maximum[10%] lifespans for lager brewing yeast strains (S.D. = standard deviation). The strains were grown on YP with maltotriose (2% w/v) agar and maintained at 25°C during the day and 4°C overnight.

Strain	Total cells manipulated	Mean lifespan	S.D.	Max. lifespan	Max.[10%] lifespan	S.D.[10%]
SCB1	106	14.56	8.84	35	29.45	3.64
SCB2	103	17.70	9.35	40	32.90	3.45
SCB3	102	17.73	6.13	34	29.50	2.92
SCB4	87	16.30	6.24	29	25.56	2.19

for SCB2 and SCB4 and decreased mean longevities for SCB1 and SCB3. The mortality profiles produced (Fig. 16.1) were very similar in each case.

In general, longevity was increased on maltose and maltotriose, and reduced on glucose indicating that medium composition significantly influences lifespan potential. This supports the observations of Barker et al.[7] that carbon source affects longevity.

Table 16.5   Mean and maximum[10%] lifespans for lager brewing yeast strains (S.D. = standard deviation). The strains were grown on YP with brewer's wort (2% w/v) agar and maintained at 25°C during the day and 4°C overnight.

Strain	Total cells manipulated	Mean lifespan	S.D.	Max. lifespan	Max.[10%] lifespan	S.D.[10%]
SCB1	77	16.97	11.37	42	36.56	3.00
SCB2	68	12.92	8.58	29	27.43	1.27
SCB3	71	16.41	8.87	36	31.50	2.62
SCB4	65	10.28	8.48	35	27.14	4.26

16.3.3   *The impact of wort on replicative senescence*

Wort represents the combination of many sugars some of which may affect longevity in *S. cerevisiae*. In addition, other constituents of this medium may influence yeast longevity. Therefore the impact of wort and batch compositional variation on yeast lifespan represents an area of significant importance to the brewing industry since this medium is utilised during both yeast propagation and fermentation. SCB1 exhibited the greatest mean and maximum[10%] lifespan on wort. The Hayflick limit, mean and maximum[10%] lifespan for strains SCB1, SCB2 and SCB4 were greater than to those observed on glucose. SCB3 demonstrated a decreased lifespan potential on wort.

Wort was observed to improve the longevity in strain SCB1 when compared with the longevity on glucose, maltose and maltotriose. Lifespan potential of strain SCB2 was reduced on wort compared with maltose and maltotriose but similar to that exhibited on glucose. The overall impact of wort on the longevity of SCB3 was negative compared with that observed on glucose and maltose, with a slight extension in longevity observed on maltotriose. The mortality profile produced by SCB4 on wort demonstrated an improved longevity when compared with glucose and a similar longevity to that observed on maltose and maltotriose. However, there are many media components which comprise wort and YP, and it is suggested that in this case, wort components other than carbon source may influence lifespan.

16.4   Conclusions

Individual cell ageing in brewing lager yeast is a strain dependent phenomenon which is primarily under genetic control. The morphological characteristics associated with progression through replicative lifespan to senescence are largely conserved, although the timing and extent to which these modifications occur are strain dependent. Environmental factors such as the nature of carbon source on which the cells are grown can also affect longevity, though the influence is not universal, with strains exhibiting specific reactions to each carbon source examined. In addition, growth on wort as opposed to YP with fermentative or respiratory carbon sources also affects yeast lifespan, though the extent and nature of this effect is strain dependant.

Deans *et al.*[15] reported a propensity for older larger cells to occur in the lower regions of the cone in conical vessels and it has been postulated that this may lead to enrichment during serial repitching of cell populations of a specific age[3,4]. It is suggested that the potential for enrichment of aged populations through serial repitching may be strain dependent and exacerbated by environmental conditions such as wort and beer composition.

It is postulated that wort batch variations in carbohydrate composition alone may influence the longevity potential of brewing yeast strains. This remains the subject of further investigation.

Acknowledgements

Dawn Rodgers is funded by the J & J Morison Scholarship and the authors would like to thank Mrs Pamela Morison-Inches for her support. The authors are also grateful to the Directors of Scottish Courage Brewing Ltd for permission to publish this work.

References

(1) Austriaco Jr., N.R. (1996) Review – To bud until death: The genetics of ageing in the yeast *Saccahromyces cerevisiae*. *Yeast* **12**(7), 623–630.
(2) Hayflick, L. (1965) The limited *in vitro* lifetime of human diploid cell strains. *Experimental Cell Research* **37**, 614–636.
(3) Smart, K.A. (1999) Ageing in brewing yeast. *Brewers Guardian* **128**, 19–24.
(4) Barker, M.G. and Smart, K.A. (1996) Morphological changes associated with the cellular ageing of a brewing yeast strain. *Journal of the American Society of Brewing Chemists* **54**(2), 121–126.
(5) Powell, C.D., Quain, D.E. and Smart, K.A., The relationship between lifespan and cellular morphology in *Saccharomyces cerevisiae*, unpublished.
(6) Sinclair, D., Mills, K. and Guarente, L. (1998) Molecular mechanisms of yeast ageing. *Annual Review of Microbiology* **52**, 533–560.
(7) Barker, M.G., Brimage, L. and Smart, K.A. (1999) Effect of Cu, Zn superoxide dismutase disruption mutation on ageing in *Saccharomyces cerevisiae*. *FEMS Letters* **177**, 199–204.
(8) Muller, I. (1971) Experiments on ageing in single cells of *Saccharomyces cerevisiae*, *Archives of Mikrobiology* **77**, 20–25.
(9) Rodgers, D.L., Kennedy, A.I., Hodgson, J.A. and Smart, K.A. (2000) Lager brewing yeast ageing and stress tolerance. *Proceedings of the European Brewery Convention Congress* **27**, 671–678.
(10) Crumplen, R.M., Slaughter, J.C. and Stewart, G.G. (1996) Characteristics of maltose transporter activity in ale and lager strains of the yeast *Saccharomyces cerevisiae*. *Letters in Applied Microbiology* **23**, 448–452.
(11) Stewart, G.G., Russell, I. and Sills, A.M. (1983) Factors that control the utilisation of wort carbohydrates by yeast. *Master Brewers of the Americas Association Technical Quarterly* **20**(1), 1–8.
(12) Gancedo, J.M. (1992) Carbon repression in yeast. *European Journal of Biochemistry* **206**, 297–313.
(13) Thevelein, J.M. (1994) Signal transduction in yeast. *Yeast* **10**, 1753–1770.
(14) Stewart, G.G. and Russell, I. (1993) Fermentation – the black box of the brewing process. *Master Brewers of the Americas Association Technical Quarterly* **30**, 159–168.
(15) Deans, K., Pinder, A., Cately, B.J. and Hodgson, J.A. (1997) Effects of cone cropping and serial repitch on the distribution of cell age in brewery yeast. *Proceedings of the European Brewing Congress* **26**, 469–476.

Part 3 Yeast Handling and Fermentation Performance

17 Yeast Handling in the Brewery

ALAN KENNEDY

Abstract Current fermentation practices generally produce two to four times the amount of yeast needed for reuse. Ideally, yeast which reproduces by asexual means should produce identical generations time and time again, and the introduction of fresh cultures should not be required. Unfortunately, yeast behaviour does change, due to a number of factors, and cultures must be replaced, normally after less than ten brewery fermentations. Yeast master cultures (optimally stored under liquid nitrogen) are used to initiate a series of laboratory and brewery 'fermentations', under aseptic conditions, in order to produce sufficient yeast for full-scale fermentation.

Achieving a consistent, quality product and allowing accurate production scheduling requires reproducible fermentations. Accurate and constant pitching rates of viable yeast cells is one of the major variables affecting fermentation performance. Pitching rates are traditionally calculated on measurements of percentage solids of yeast cells in a slurry, in conjunction with the percentage viability measured by vital staining techniques. More recently, automatic pitching systems based on capacitance measurements of yeast suspensions have been used. Estimations of yeast vitality can also be made. Following fermentation, it is critical to future fermentation performance that the brewer optimizes yeast cropping procedures to ensure the integrity (viability and vitality) of the production yeast strain.

17.1 Yeast husbandry and propagation

The main aims of yeast husbandry and propagation are to achieve consistent process and product quality, and a number of areas need to be considered.

17.1.1 *Yeast strain selection*

Selection of a strain may be from a commercial culture collection according to description. Alternatively, the strain may be defined by franchise requirements, in which case the franchisor's laboratory maintains and issues the yeast. Often strains are selected, or rather re-selected, from existing pitching yeast.

Several incentives for strain selection or re-selection can be proposed. Conversion from open to closed conical vessel may require a bottom cropping yeast. New products may require new flavours. A desire to improve process efficiency may result in selection for strains which can ferment faster, ferment to a lower gravity or give reduced growth or produce less fobbing during fermentation (better percent vessel utilisation). If a strain has become unstable (most commonly seen with flocculation and sedimentation behaviour) then performance variability and loss of process control require correction by re-selection. A desire to change from production using mixed strains necessitates selection of a single yeast with suitable flavour and process characteristics. If a brand volume decreases below a certain percentage of production then replacement with another strain currently in use at the brewery may become necessary.

17.1.2   *Storage and maintenance of yeast*

Preservation procedures commonly employed include agar slopes, deep freezing, freeze drying and storage under liquid nitrogen. Whatever the method of preservation, yeast slopes are normally prepared for the initiation of a new yeast culture. A single use of yeast slopes is recommended rather than repeated subculture. Slopes should be made in batches, according to numbers required. Central production of batches of slopes for distribution to breweries is commonly practised. For each batch of slopes prepared, at least one should be utilised for testing and two or three retained for reference testing, in case of any problems that are encountered during use at the breweries. In addition, yeast karyotyping techniques (such as CHEF) can be used to confirm the identity of a second test slope prior to dispatch.

17.1.3   *Yeast propagation*

Upon repeated use (re-pitching in the brewery) it may be observed that desirable characteristics are lost or become modified through selective changes in the distribution of yeast types in a population. Trub may accumulate, especially in highly hopped, bottom cropped fermentations. Periodic re-establishment of purity due to contamination with bacteria or wild yeast may be necessary. Some yeast may show a reduction in vigour after a long period of use, especially in high gravity fermentations ($> 14°$Plato). A suitable propagation regime must give sufficient quantity of pure, fermentatively active yeast in a planned time period.

Conventionally, as the propagation process advances there is a stepped increase in volume and a reduction in aeration rates and growth temperatures to those in production fermentations. A typical propagation regime would include a number of steps, for example: (i) yeast slope; (ii) 10 ml wort/medium in flask; (iii) 1000 ml wort/ medium in flask; (iv) 10 to 40 litres of wort in a Carlsberg flask (or converted keg); (v) culture plant stage 1 (15 to 25 hectolitres); and (vi) culture plant stage 2 (75 to 100 hectolitres).

Step volumes and the number of steps depend on the quantity of wort to be pitched in the first production brew. In some cases, the yeast from the second stage of the culture plant will have to be pitched into a half brew before sufficient yeast is available for a full brew. Parameters monitored during propagation in a culture plant vary, the most common include temperature, rate of aeration, gravity drop, yeast cell count and viability, ergosterol content and inspection for microbiological contamination. High ergosterol level in the cultured yeast shows that delivery of oxygen has been sufficient. Transfer criteria (such as time, cell count, biomass and gravity) are aimed at ensuring that the transferred yeast has achieved adequate growth, and is still in an actively growing mode (exponential growth phase). The time taken from starter culture to new pitching yeast is strain and temperature dependent. The first production brew is sometimes considered as part of the propagation regime. A range of 7 to 21 days may be encountered.

Microbiological contamination is a risk at all stages of propagation, and must be strictly controlled, especially during the early stages. In most propagation regimes, the detection of microbiological contamination at any stage should result in the culture being aborted, and propagation re-started from a fresh slope[1].

Inefficient transfer of dissolved oxygen is a common reason for slow propagations with poor yields. If a high growth temperature is used throughout the propagation, then poor adaptation to lower pitching and fermentation temperatures may result. If transfer of yeast to subsequent propagation steps is too soon, then poor yield and long lag times will result. If transfer is too late, then the longer lag times resulting from the yeast having entered stationary phase will prolong the entire process.

17.2   Good practice in yeast handling

17.2.1   *Functional requirements*

Brewing yeast must be cultured, cropped and stored in such a way as to ensure: correct strain; phenotypical homogeneity (flocculence, metabolism, age); freedom from infection by wild yeast, other brewing yeasts and bacteria; high viability; and consistent solids to barm ratio. To this end the physical requirements of the storage system would include: effective cold storage (3°C); effective agitation for homogeneity; atmospheric pressure; effective cleaning and sterilisation; yeast sanitisation (acid washing); microbiological integrity; and compatible capacities.

17.2.2   *Operational requirements*

17.2.2.1   *Culturing.*   As stated above, yeast culturing must provide phenotypically homogeneous yeast of the correct strain, free from infection and in sufficient quantity and health to effect normal fermentation in generation 1. The yeast generation number must not exceed the specification for each strain. Introduction of new cultures (onsets) into full-scale production must be managed in such a way as to ensure regular replacement of old with new. There should not be more than two onsets in use at any time. Old and new onsets must never be mixed.

17.2.2.2   *Cropping from fermenter for reuse.*   Yeast for repitching must be cropped only from fermentations which meet specification for attenuation profile and VDK reduction. Wherever possible, the microbiological status of fermentations should be available in time to prevent re-use of infected or cross-contaminated yeast. Typically, yeast cropping will be initiated within 24 hours of the donor fermentation reaching target coolback temperature[2]. The yeast cropping operation will usually include diversion of trub and first yeast portion to waste. Yeast crops must not be mixed into storage. Yeast cropping procedures must be optimised to target a consistent ratio of yeast to barm (% solids) as possible. Yeast in a cone at the end of fermentation should not just be considered as a homogeneous mass. It is believed that a gradient of yeast quality exists in the cone fraction, so yeast cropping procedures (top and tail) may have a significant effect on subsequent fermentations[1]. Each yeast strain has a limit to the number of divisions that it can go through before 'yeast ageing' (senescence) effects become evident. The cellular lifespan of some yeasts can be as low as 8 divisions. Senescence in a population can be seen as a slow down in replication and metabolism, accompanied by the appearance of deficiencies in the yeast[3]. Older cells (in cellular lifespan terms) are significantly larger than younger, more 'vital' cells.

17.2.2.3 *Pitching into a new fermentation.* Dedicated pitching tanks sized for the pitching of a single fermentation are usually included in any new installations and acid washing is effected in the pitching tanks. Pitching rate control may be facilitated by weighing out (load cells) a batch of yeast correctly calculated to match the wort volume to be pitched. Alternatively, instruments such as the Aber meter may be used[4]. Many brewers believe that under pitching is potentially more serious than over pitching. A typical target pitching rate would be approximately 33% higher than the rate below which the fermentation profile becomes irregular (e.g. slow to attenuate, high VDK). Pitching yeast viability should be 90% (by methylene blue staining) before and after acid washing. The pitching rate may not be adjusted for viability because of the risks associated with adding excessive non-viable cells.

17.2.2.4 *Management considerations.* In a modern brewery, software control is exploited wherever possible to ensure correct procedures regarding yeast movements. This could include operational inhibits such as preventing storage tanks being filled with yeast until CIP'd, and pitching the wrong yeast strain into specified wort qualities. Line purges between different yeast strains are automated where possible. Records relating to yeast history, quality and pitching rate will be computer generated with minimum manual input and calculation wherever possible. All records relating to yeast movements, hand-written or computer-generated, must be legible and fully traceable/auditable, forming part of the company 'quality system'.

Yeast will be managed as a raw material, with pitching rate quantities, sources and destinations planned well ahead of cone removal. The transfer of yeast from storage tanks to waste should be a rare requirement except where existing tanks are inappropriately sized for fermenter requirements, and small residual quantities are not required for fermentation. Transport of yeast slurries between breweries must be kept to a minimum. The possible requirement to do so, however, must be considered when designing yeast handling plant.

17.2.2.5 *Contents and pitching quantity measurement.* Storage and pitching vessels will be fitted with load cells, and pitching quantity will be calculated on weight as recorded by the load cells, with % solids measured off-line, and expected wort volume calculated. Yeast viability will not be accounted for in the calculation but will always require off-line measurement in order to affirm the suitability of the yeast for reuse.

17.3 Yeast viability and vitality

17.3.1 *Yeast viability*

Yeast viability is a measure of the number of living cells in a population. More accurately, it has been described as the ability of cells to grow, reproduce and interact with their immediate environment. Various methods can be used to estimate yeast viability. Monitoring cell viability via plating and slide cultures is based on cell growth, while staining procedures are based on cell damage and/or metabolic activity. Methylene blue vital staining remains the industry standard for estimating yeast viability. This brightfield staining procedure involves the apparent exclusion or reduction of the methylene blue dye by viable cells. There are now concerns over the accuracy of the test, especially that the technique may overestimate the level of living cells in a population of low viability. Recent studies[5] have suggested that methylene violet may be a more suitable alternative to methylene blue. It is reported that the dye is less pH dependent and is accurate across the full range of viabilities. In addition, there is reportedly no gradation of colour between cells, which may be seen with methylene blue stains.

17.3.2 *Yeast vitality*

The vitality of a yeast population has variously been described as a measure of activity, fermentation performance or the capacity to overcome and recover from physiological stress. It may be best to consider vitality a combination of all these factors.

17.3.2.1 *Vitality measurement.* Many methods have been suggested for the measurement of yeast vitality, based in a number of areas: (1) metabolic activity; (2) cellular components; (3) fermentation capacity; (4) acidification power; (5) oxygen uptake ability; and (6) other methods. Further, specific details of some of these methods can be found in review articles[6,7]. As mentioned above, many factors relating to how yeast is handled in the brewery can affect the yeast's vitality (and viability)[8].

17.3.2.2 *Potential problems.* A number of potential problems exist with yeast vitality testing in the production environment.

- Method used must be calibrated against some aspect of fermentation performance
- Yeast vitality is only one of many factors affecting fermentation consistency
- Correlation has been shown between vitality test and laboratory fermentation, but not on production scale
- There are many causes of loss of vitality, but each method measures only a single parameter
- The majority of tests available are laboratory based and time consuming
- No in-line instrumentation is currently available.

17.3.2.3 *Use of yeast vitality tests.* Some brewers do not rely on methylene blue staining alone for yeast viability and vitality testing. Work reported by South African Breweries[9] indicates that they use a battery of three different tests on yeast samples from a well mixed storage tank prior to pitching. Two tests give an estimation of yeast viability while a third measures vitality. The measurement of the pH of the supernatant from a centrifuged sample of yeast relates to viability, i.e. if the pH of the slurry is higher than that of the final beer then yeast death and autolysis must have taken place. Similarly, the measurement of protease release from the same sample provides a more sensitive measure of cell death and autolysis. The third test, a magnesium release test, uses a commercially available test kit to measure Mg^{2+} ion release from the yeast sample. SAB report that vital yeast will release magnesium immediately on contact with production wort. A specification is set for each test, and it is reported that these tests are more predictive of fermentation performance than any others investigated.

17.4   Conclusions

While an accurate and efficient measurement of yeast vitality would provide the brewer with useful information on the condition of the yeast prior to pitching, no such in-line test is yet available. A number of different methods for testing yeast vitality have been suggested, but each has its own problems. By ensuring that defined yeast management procedures are followed at all times, that yeast viabilities remain above 95% and that accurate yeast pitching control is achieved, then an additional vitality measurement may not be needed. Some brewers, though, do make use of vitality testing routinely.

Acknowledgement

The author wishes to thank the directors of Scottish Courage Brewing Ltd for permission to publish this article.

References

(1)   Jones, H.L. (1997) Yeast propagation – past, present and future. *Brewer's Guardian* **126**, 24–27.
(2)   O'Connor-Cox, E (1997) Improving yeast handling in the brewery. Part 1: Yeast cropping. *Brewer's Guardian* **126**, 26–34.
(3)   Smart, K.A. (1999) Ageing in brewing yeast. *Brewer's Guardian* **128**, 19–24.
(4)   Carvell, J. (1994) Automatic yeast pitching control using a yeast sensor. *The Brewer* **80**, 57–59.
(5)   Smart, K.A., Chambers, K.M., Lambert, I., Jenkins, C. and Smart, C.A. (1999) Use of methylene violet staining procedures to determine yeast viabilty and vitality. *Journal of the American Society of Brewing Chemists* **57**, 18–23.
(6)   Lentini, A. (1993) A review of the various methods available for monitoring the physiological status of yeast: yeast viability and vitality. *Ferment* **6**, 321–327.
(7)   Boulton, C.A. (1996) A critical assessment of yeast vitality testing. *Ferment* **9**, 222–226.
(8)   Edelen, C.L., Miller, J.L. and Patino, H. (1996) Effects of yeast pitch rates on fermentation performance and beer quality. *MBAA Tech. Quart.* **33**, 30–32.
(9)   O'Connor-Cox, E. (1998) Improving yeast handling in the brewery. Part 3: Yeast pitching and measurement of yeast quality. *Brewer's Guardian* **127**, 20–26.

18 Yeast Pitching with Relatively High Variability in Yeast Slurry Concentration

C. HOLMES and H.A. TEASS

Abstract   The typical yeast pitching method depends upon a slurry volume concentration being blended into the work line. Cell counting is by laboratory analysis or use of an on-line live cell counter calibrated by the laboratory. Alternatively, there is this new method which directly measures the yeast slurry. This automatic haemocytometer counter method incorporates a new optical electronic technology: calculating cell number by measurement of slurry optical density. This is an automatic method that has been operating successfully in a brewery. The effectiveness of pitching has been observed and there has been significant improvement over the older volume laboratory methods, resulting in enhanced beer uniformity. Data provided here demonstrate pitching improvements. This automatic system may be calibrated to a haemocytometer as such or a haemocytometer adjusted for live cell percentage (by methylene blue viability analysis). In either case there is a substantial increase in speed and simplicity in yeast concentration analysis. With this device there is substantial cost saving over the alternative methods of analyses and the cell volume error effect is minimised. It comes packaged to include the cell counter and flow meter. It is inert to cleaning agents and is suitable for in-line cleaning.

19 Detection of Lactic Acid Bacteria in Pitching Yeast Slurries Using PCR

BENHAM TAIDI and JEFF HODGSON

Abstract   The rapid detection of two previously isolated brewery contaminant organisms in yeast slurries was optimised using PCR. The detection of bacterial DNA was carried out against a background of yeast DNA. Two DNA isolation methods, namely chemical and enzymatic, were tested for DNA yield. Chemical isolation of DNA gave a higher yield of DNA than the enzymatic method. Detection of brewery isolated lactobacilli and a *Pediococcus* strain was possible using the primers 907r and LB-95, which yielded a characteristic PCR product of 900 bp length. The PCR reactions were very reproducible but many additional fragments were also obtained from these which were attributed to non-specific DNA and primer interactions. Hence, detection of *Lactobacillus* and *Pediococcus* DNA in a background of brewing yeast DNA was possible.

19.1   Introduction

Yeast handling best practice stipulates storage of yeast pitching slurries for as short a period as possible and in any case no longer than 48 h. This period of time is too short to allow actionable microbiological examination of the yeast slurry with traditional microbiological methods. The rapid rate of yeast reuse in the brewing industry has increased the necessity for the development and use of rapid microbiological techniques for the detection of contaminating organisms in yeast slurries.

Many rapid microbiological techniques are now available with a variety of applications. The polymerase chain reaction (PCR)[1] is one such tool which can be used for the detection and identification of micro-organisms potentially present within food as contaminants[2-6] and also for confirmation of source organisms used for food manufacture[7-9]. Targeted PCR relies on the amplification and detection of species-specific sequences of DNA. These sequences are then used as indicators for the presence of an organism or a group of organisms.

PCR has two immediate potential applications within the modern brewing industry, namely yeast strain confirmation and detection of microbiological contamination in pitching yeast. The production of a large variety of brands at the same brewery can necessitate the handling of a number of yeast strains. For quality assurance purposes it is important to confirm the identity of yeast strains periodically. A rapid fingerprinting technique such as PCR is ideally suited to such applications but still requires to be refined to a degree capable of differentiating very closely related strains. Lager yeasts are thought to have a very close genetic origin, making them difficult to differentiate[10]. The second application of PCR in the brewing industry is the rapid detection of contaminating micro-organisms such as wild yeast, lactic acid bacteria and acetic acid bacteria. Rapid detection of contaminant organisms in a yeast slurry is very important because yeast in the brewery is recycled many times. Recycling of pitching yeast necessitates carefully defined yeast handling practices[11], part of which is the specification of very short (less than 48 hours) storage periods prior to reuse. The ability to detect contaminant organisms during the relatively short period of yeast storage would provide a potent tool for quality control of pitching yeast slurries. The detection method of choice should not be too specific as this would slow down the analysis. Ideally, detection of a wide variety of contaminating organisms would be carried out with only a few PCR reactions. Careful identification of the organisms can be made subsequently if necessary.

This chapter describes a trial of two DNA extraction techniques and the use of several published[4,5] primers for the detection of brewery isolates of lactic acid bacteria. The optimised procedure was then used to detect bacteria in an artificially contaminated yeast slurry. Future work will concentrate on testing the sensitivity of the method against a wider range of lactic acid bacteria and to decrease the time taken for the analysis.

19.2   Materials and methods

Brewery isolates *Lactobacillus paracasei paracasei* and *Pediococcus pentosaceus* were grown using Raka Ray broth (Difco). These isolates originated from beer forcing samples and were identified using appropriate API strips. The bacterial cultures were incubated at 27°C for 48 h in order to obtain a culture for DNA extraction.

DNA extraction from bacteria and yeast was carried out either by phenol/chloroform (1:1 v/v) in the presence of glass beads (based on Yasui *et al.*[5]) or by an enzymatic method (Puregene, Gentra). PCR reactions were performed using Ready-To-Go (Pharmacia) PCR beads with additional magnesium (3.5 mM $MgCl_2$ final concentration). For PCR reactions an initial period of 4 min at 95°C was followed by 45 cycles of 1 min at 95°C, 1 min at 58°C and 1 min at 72°C. PCR products were stained with ethidium bromide and visualised on agarose (1.3% w/v) gels. The primers used to detect lactic acid bacteria were based on 16S rRNA sequences (Table 19.1).

Table 19.1   PCR primers used.

Name of primer	Sequence of primer	Reference
907r	5′CCGTCAATTCCTTTGAGTTT3′	5
LB-95	5′AAGTCGAACGAGCTTCC3′	5
DA-40	5′GTCTCCTAACTGATAGCT3′	5
LbHC-1	5′ATCCGGCGGTGGCAAATCA3′	4
LBHC-2	5′AATCGCCAATCGTTGGCG3′	4

19.3   Results and discussion

Two DNA extraction methods were tested on 48 h cultures of three brewery isolates previously characterised as *Lactobacillus paracasei paracasei* and another isolate identified as *Pediococcus pentosaceus*. The quantity of DNA obtained by mechanical means was estimated visually, based on the intensity of the bands on an agarose gel, to have been greater than that obtained by enzymatic extraction (results not shown). Different combinations of the primers were tested with DNA extracted from pure bacterial cultures. The base pair Lb HC-1 and Lb HC-2 resulted in no reactions. The base pair 907r and DA-40 gave very weak reactions. The base pair 907r and LB-95 gave the best reaction with brightest 900 bp bands. According to Yasui et al.[5] the presence of a 900 bp PCR product indicated the presence of L. lindneri. The primers 907r and LB-95 were also suitable for the detection of the brewery isolates of lacto-bacilli and Pediococcus.

The reproducibility of the PCR reaction employing the primers 907r and LB-95 was tested with both mechanically and enzymatically extracted DNA (Plate 7). A 900 bp band was consistently and reproducibly obtained with the DNA from the four brewery isolates; however, many bands of variable molecular weight were also obtained. These bands are thought to be non-specific PCR products.

DNA was extracted from a sample of brewery yeast slurry shown to be con-taminated with lactic acid bacteria. The primer pairs 907r and LB-95 were used to carry out PCR reactions with the extracted DNA, which contained both bacterial and yeast DNA. The presence of lactic acid bacteria was confirmed in the sample by the appearance of a distinctive band at 900 bp. The level of bacterial contamination in the yeast slurry was not known, but the ratio of yeast DNA to bacterial DNA is expected to have been high. The yeast DNA did not interfere with the PCR reaction between the primer set 907r and LB-95 and bacterial DNA. Using a specific set of primers it was also possible to confirm the identity of the yeast strain in a separate PCR reaction without interference from the bacterial DNA (results not shown). This demonstrated that the method could be applied to potentially contaminated yeast slurries. The procedure currently requires two working days to complete.

Acknowledgement

The authors wish to thank the directors of Scottish Courage Brewing Ltd for per-mission to publish this research.

References

(1)   Mullis, K. and Faloona, F.A. (1987) Specific synthesis of DNA in vitro via a polymerase-catalyzed chain reaction. Methods in Enzymology **155**, 335–350.

(2)   Tsuchiya, Y., Kaneda, H., Kano, Y. and Koshino, S. (1992) Detection of beer spoilage organisms by polymerase chain reaction technology. J. Amer. Soc. Brew. Chem. **50**, 64–67.

(3)   DiMichele, L.J. and Lewis, M.J. (1993) Rapid, species-specific detection of lactic acid bacteria from beer using the polymerase chain reaction. J. Amer. Soc. Brew. Chem. **51**, 63–66.

(4)   Sami, M. and Yamashita, H. (1997) A new rapid method for determination of beer-spoilage ability of lactobacilli. J. Amer. Soc. Brew. Chem. **55**, 137–140.

(5)   Yasui, T. Okamoto, T. and Taguchi, H. (1997) A specific oligonucleotide primer for the rapid detection of Lactobacillus lindneri by polymerase chain reaction. Can. J. Microbiol. **43**, 157–163.

(6)   Sami, M., Yamashita, H., Hirono, T., Kodokura, H., Kitamoto, K., Yoda, K. and Yamasaki, M. (1997) Hop-resistant Lactobacillus brevis contains a novel plasmid harbouring a multidrug resistant-like gene. Journal of Fermentation and Bioengineering **84**, 1–6.

(7)   de Barros Lopes, M., Soden, A., Henschke, P.A. and Langridge, P. (1996) PCR differentiation of

commercial yeast strains using intron splice site primers. *Applied and Environmental Microbiology*, December, 4514–4520.

(8) Tompkins, T.A., Stewart, R., Savard, L., Russell, I. and Dowhanick, T.M. (1996) RAPD-PCR characterisation of brewing yeast and beer spoilage bacteria. *J. Amer. Soc. Brew. Chem.* **54**, 91–96.

(9) Laidlaw, L., Tompkins, T.A., Savard, L. and Downhanick, T.M. (1996) Identification and differentiation of brewing yeasts using specific and RAPD polymerase chain reaction. *J. Amer. Soc. Brew. Chem.* **54**, 97–102.

(10) Hammond, J.R.M. (1996) Yeast genetics. In *Brewing Microbiology*, Priest, F.G. and Campbell, I. (eds), Chapman and Hall, London, pp. 43–82.

(11) O'Connor-Cox, E. (1998) Improving yeast handling in the brewery. Part 2: Yeast collection. *Brewer's Guardian*, February, 22–34.

20 Yeast Oxygenation and Storage

CHRIS BOULTON, VINCENT CLUTTERBUCK and SEAN DURNIN

Abstract The practice of serial fermentation coupled with intervening cropping and storage of brewing yeast produces a requirement for wort oxygenation and creates an opportunity for inconsistency in fermentation performance to arise. The yeast sterol pool synthesised during the aerobic phase of fermentation is diluted by subsequent cell proliferation until lack of membrane function produced by sterol deficiency becomes one of the factors that restrict further growth. In this respect, the concentration of oxygen supplied at the start of fermentation regulates the quantity of yeast sterol synthesised and, by inference, the extent of subsequent yeast growth. However, should yeast be exposed to oxygen during the storage phase between cropping and re-pitching, some limited sterol synthesis can occur at the expense of the dissimilation of glycogen reserves. In such a case, in the next fermentation it would be necessary to reduce the wort dissolved oxygen tension to allow for the high initial sterol concentration in the pitching yeast. Failure to take this preventive step will result in excessive yeast growth during fermentation, with a concomitant reduction in efficiency and the possibility of unacceptable changes in the concentrations of flavour-active metabolites dependent on yeast growth.

This chapter describes an alternative strategy in which yeast is deliberately exposed to oxygen in a controlled process, prior to pitching. This produces pitching yeast, which is replete in sterol and thereby ostensibly removes the requirement for subsequent wort oxygenation. In so doing fermentation control is simplified since selection of an appropriate pitching rate becomes the only variable used to regulate the extent of yeast growth. The experience gained of the application of this strategy at production scale is described and critically discussed.

20.1 Introduction

Conventional fermentation control regimes require initial selection and regulation of the pitching rate and wort dissolved oxygen concentration. Subsequently, fermentation rate is controlled by the application of an appropriate attemperation profile. Such approaches do not correct for variability in the physiological condition of the pitching yeast. Thus, if pitching yeast is exposed to oxygen during storage some limited sterol synthesis may occur. Theoretically, such yeast would have a reduced requirement for wort oxygenation in order to produce standard fermentation performance[1]. Conversely, yeast stored at an inappropriate elevated temperature, or for a prolonged period of time, but with no exposure to oxygen, would have a stressed, starved physiology, typically with reduced glycogen reserves and low sterol content. In order to achieve standard fermentation performance with such yeast it would be necessary to increase the pitching rate and possibly the wort oxygen concentration.

The ability of yeast to synthesise sterol under conditions of non-growth in the absence of wort but in the presence of oxygen may be utilised in a novel fermentation control strategy. In this approach pitching yeast is forcibly exposed to oxygen in a controlled process[2]. This allows sterol synthesis to proceed to a similar extent to that seen in an oxygenated wort fermentation. Consequently, the requirement for wort oxygenation is removed and the desired fermentation profile may be achieved simply by the selection of an appropriate pitching rate. Implicit in this strategy is the assumption that oxygenated yeast should have a more consistent physiology compared with conventional pitching yeast. Results of a study reported elsewhere have tentatively confirmed this premise[3].

Data presented here illustrate some of the biochemical changes that occur during the oxygenation of brewery pitching yeast. The design and operation of plant suitable for oxygenating yeast at production scale is described, and the results of production-scale trials are discussed.

20.2   Materials and methods

Yeast was obtained from brewery storage vessels. Laboratory yeast oxygenation was performed at room temperature using pitching yeast slurries (20 to 45% wet. wt/vol.) suspended in beer. Slurries (2 litres) were contained within a 5 litre conical flask and stirred magnetically to improve gas transfer rates. Oxygen was delivered to the slurry via a glass sinter from a cylinder fitted with a pressure reducing valve and rotameter. Oxygen uptake rates were measured by discontinuing the gas supply and monitoring the decrease in oxygen tension using a polarographic oxygen electrode. Sterols were determined as trimethylsilyl derivatives using capillary GC and cholestan-β-ol as internal standard. Glycogen was determined enzymatically[8].

20.3   Results and discussion

The effects on the intracellular concentrations of total sterol and glycogen when pitching yeast, suspended in beer, was oxygenated are shown in Fig. 20.1. The total

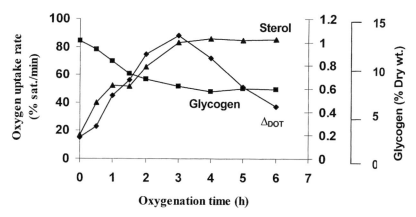

Fig. 20.1   Effects on the intracellular concentrations of glycogen and total sterol, compared with the rate at which the yeast took up oxygen from the suspending medium, when a pitching yeast slurry (35% wet wt/vol., suspended in beer) was oxygenated at 20°C.

sterol content increased from an initial value of approximately 0.2% to just over 1% of the cell dry weight after 3 to 4 h oxygenation. Thereafter, there was no further increase in the sterol content of the yeast. The quantity of sterol synthesised was of a similar magnitude to that which occurs in the aerobic phase of fermentation[4]. The increase in sterol content was accompanied by a concomitant decrease in the intra-cellular concentration of glycogen. This was in accord with the contention that gly-cogen dissimilation provides energy for sterol synthesis in brewery fermentation[3]. During the oxygenation treatment there was an increase in the rate at which the yeast took up oxygen. The maximum sterol and minimum glycogen concentrations coin-cided with the maximum observed yeast oxygen uptake rate, and thereafter the latter parameter decreased.

The data in Fig. 20.2 demonstrate the fermentation performance of oxygenated yeast. In this experiment samples of yeast were removed from the oxygenation vessel, after the treatment times indicated, and pitched at the same viable rate into anaerobic wort contained in EBC tall-tubes. Fermentation rates increased progressively with the duration of oxygenation, up to the time at which the maximum oxygen uptake rate was observed. Yeast removed at and after the maximum observed oxygen uptake rate produced essentially identical fermentation performance. This demonstrates the essence of the fermentation control strategy.

During the oxygenation process the pH of the suspending beer decreased from an initial value of approximately 4.2 to just under 3.8 at a time coincident with the maximum observed oxygen uptake rate. The yeast dry weight also decreased during

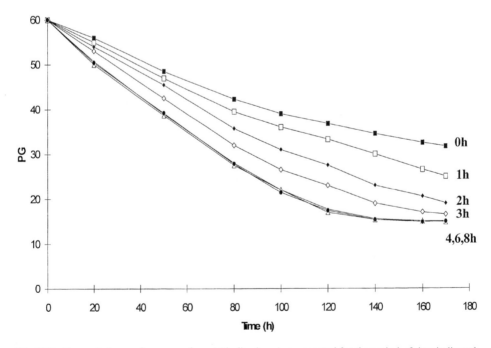

Fig. 20.2 Fermentation performance of yeast (decline in pg), oxygenated for the period of time indicated, pitched into EBC tall-tubes containing 2 litres of 12°Plato lager wort attemperated at 12°C.

the oxygenation process, presumably a consequence of glycogen dissimilation (Fig. 20.3(a)). Some of the carbon generated by glycogen breakdown was apparently metabolised to ethanol, and during the oxygenation process the concentration of the latter increased by approximately 1.5 g/l. There was no change in yeast viability during oxygenation (Fig. 20.3(b)).

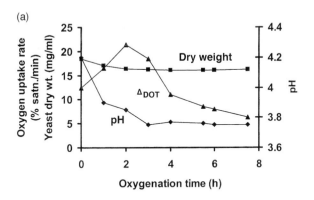

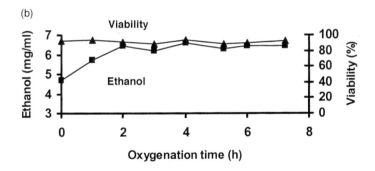

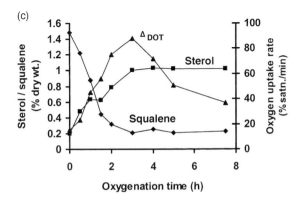

Fig. 20.3 Changes in (a) pH/yeast dry weight, (b) yeast viability/ethanol and (c) sterol/squalene compared with changes in the rate of yeast oxygen uptake during oxygenation of a lager pitching yeast slurry.

The formation of sterol during oxygenation was mirrored by a decrease in the intracellular squalene pool (Fig. 20.3(c)). Evidently the quantity of sterol formed could be entirely accounted for by cyclisation of the squalene already present in the yeast at the onset of oxygenation. The principal sterols formed were ergosterol and zymosterol (Fig. 20.4(a)). However, in relative terms the proportion of ergosterol decreased in inverse proportion to zymosterol (Fig. 20.4(b)).

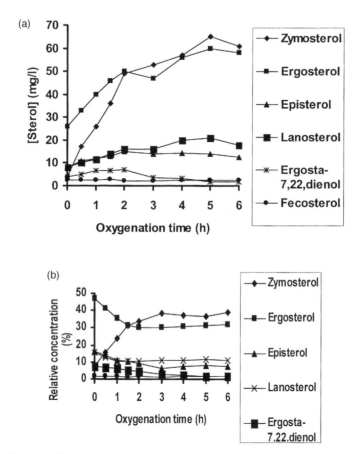

Fig. 20.4 Changes in the concentrations of (a) individual sterols and (b) the relative proportions of each sterol during oxygenation of a lager pitching yeast slurry.

The effects of prolonged oxygenation are shown in Fig. 20.5. The initial peak of oxygen uptake rate was followed by a second larger peak which occurred after some 16 to 24 hours oxygenation. The second increase in oxygen uptake rate coincided with a gradual disappearance of exogenous ethanol. Presumably this indicated that after 16 h treatment the yeast had become fully respiratory and, therefore, capable of oxidising ethanol. However, the majority of the sterol synthesis occurred during the first 4 h of oxygenation, there being only a modest increase after this time. Thus, even after prolonged oxygenation and induction of respiratory pathways the sterol con-

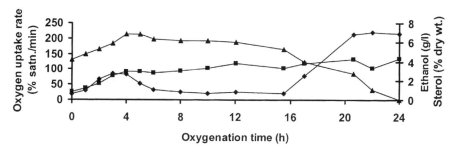

Fig. 20.5 Effect of prolonged oxygenation of a lager pitching yeast slurry on the concentrations of ethanol and sterol compared with the rate of yeast oxygen uptake.

centration did not increase to the relatively high levels (5% cell dry weight) associated with derepressed yeast[5].

Plant suitable for yeast oxygenation at production scale is shown in Fig. 20.6. It consisted of a hygienically designed stainless steel process tank with an operating volume of 80 hl, jacketed for rapid cooling and mounted on load cells. Pure oxygen was supplied via a thermal mass flow meter through a stainless steel candle. High rates of oxygen transfer were ensured by provision of a powerful mechanical agitator. A constant dissolved oxygen tension was achieved by taking output from a dissolved oxygen probe and using this to control the rate of oxygen addition. The temperature of the yeast slurry was controlled by circulation through an external loop system

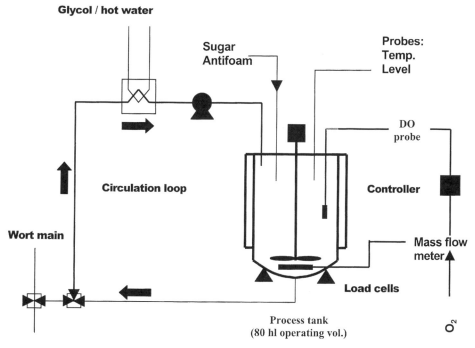

Fig. 20.6 Schematic of plant used for yeast oxygenation on a production scale.

fitted with a plate and frame heat exchanger. The latter was provided with facilities for cooling and heating.

Yeast slurry suspended in barm ale (20 to 45% wet wt/vol.) was transferred to the process tank immediately after cropping from fermenter and the temperature increased to 20°C by circulation through the heat exchanger. During the oxygenation process the temperature was maintained at 20°C and the dissolved oxygen tension (DOT) at a set point roughly equivalent to air saturation (5 to 8 ppm). In order to compensate for yeast with a low glycogen content the slurry was supplemented with maltose syrup at a final concentration of 3% (w/v). The process was considered complete when the maximum oxygen uptake rate was achieved. This was inferred from the demand for oxygen needed to obtain a constant DOT. When oxygenation was finished the yeast was either pitched immediately or, if not required until later, cooled to 3°C for storage.

The results of an oxygenation trial using a production lager yeast strain are shown in Fig. 20.7. With this yeast the maximum oxygen uptake rate occurred after about 2 h treatment, as judged by the oxygen flow rate. This was accompanied by a transient peak of exothermy, based on the flow of coolant required to maintain the set-point temperature. The added maltose disappeared rapidly from the slurry simultaneously with dissimilation of yeast glycogen. The utilisation of these carbon sources was accompanied by a transient decrease in pH and formation of ethanol and sterol. Possibly the pattern of ethanol formation was biphasic, indicating that some derived from glycogen and some from the added maltose. As with the laboratory trials, there was a decrease in the yeast dry weight and there was no significant change in viability, which remained high throughout the treatment.

In the majority of the brewery trials it was necessary to store the oxygenated yeast for a period of up to 48 h, prior to pitching into anaerobic wort. Providing normal storage temperatures were used (2 to 4°C) the oxygenated yeast withstood relatively long periods of starvation as well as normal pitching yeast. At elevated temperatures, however, standard pitching yeast retained high viability during storage for longer periods compared with oxygenated yeast (Fig. 20.8). Presumably this was a consequence of the oxygenated yeast having little glycogen reserves to provide maintenance energy during storage.

The fermentation performance of oxygenated yeast, at production scale, compared with control fermentations using similar unoxygenated yeast, is shown in Fig. 20.9. There was a demonstrable improvement in the consistency of fermentation profiles. However, in order to achieve similar vessel turn-round times it was necessary to increase the pitching rates of the oxygenated yeast fermentations by 30% compared with the wort oxygenated controls.

Beers from oxygenated yeast fermentations were considered true-to-type using standard flavour profiling. In quantitative terms the concentrations of total esters and higher alcohols were within specification and there appeared to be greater consistency in the beers derived from yeast oxygenated fermentations (Fig. 20.10). Similarly, the total yeast growth was more consistent in the trial fermentations (Fig. 20.11). However, as may also be seen, the extent of yeast growth in the trial fermentations was significantly reduced compared with controls. Surprisingly, in separate laboratory trials it was demonstrated that a combination of yeast oxyge-

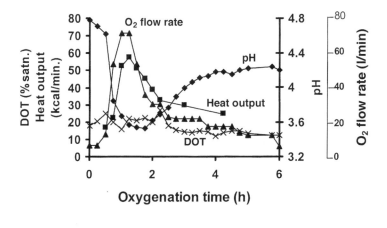

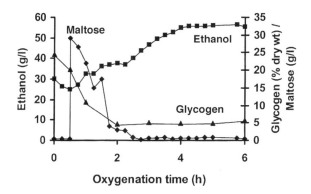

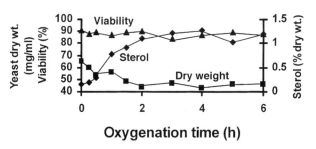

Fig. 20.7 Changes occurring during the oxygenation of lager pitching yeast slurry using the plant shown in Fig. 20.6. The slurry contained approximately 20% wet weight yeast. Maltose syrup to give a final concentration of 3% was added after 30 min. During oxygenation the temperature was maintained at 20°C.

nation and oxygenated wort did not greatly increase yeast growth during fermentation (Fig. 20.12). Thus, in the presence of 18 ppm wort dissolved oxygen, the oxygenated yeast fermentation produced a crop in which the new growth was equivalent to approximately 12 g/l wet wt yeast compared with new growth of 17

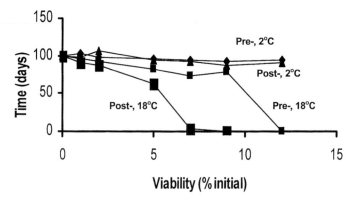

Fig. 20.8 Changes in viability with storage of pitching yeast pre and post oxygenation. Slurries were stirred continually during storage under a nitrogen atmosphere.

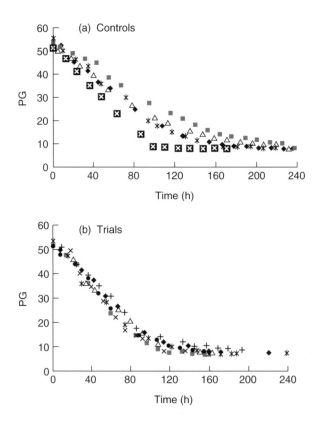

Fig. 20.9 Fermentation profiles (1600 hl, 12°Plato lager type) using (a) standard pitching yeast and oxygenated wort (controls) and (b) oxygenated yeast and anaerobic wort (trials).

Total esters

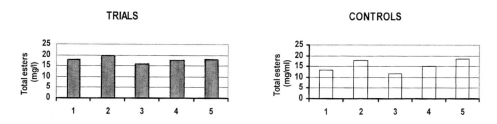

Total higher alcohols

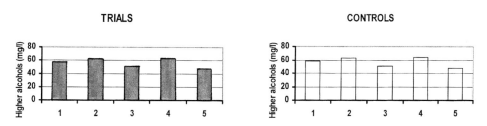

Fig. 20.10 Total ester and higher alcohol content of trial and control beers.

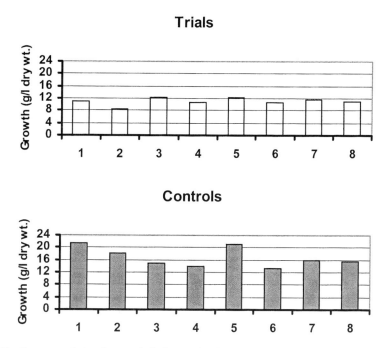

Fig. 20.11 Yeast growth (total crop, pitched yeast) in trial and control fermentations.

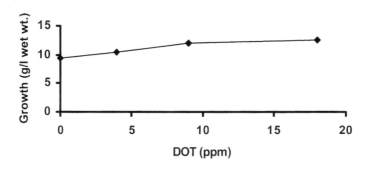

Fig. 20.12   Effect on yeast growth (total crop, pitched yeast) of using combinations of wort and oxygenated yeast. Fermentations were performed in EBC tall-tubes.

to 21 g/l wet wt in the case of a control fermentation using standard pitching yeast and 18 ppm wort oxygenation.

20.4   Conclusions

It has been demonstrated that yeast, suspended in beer, cropped from fermenter and forcibly exposed to oxygen, will accumulate sterol to a similar concentration to that which is synthesised by yeast during the aerobic phase of a normal oxygenated wort fermentation. The oxygenation process was accompanied by a transient increase in the rate at which the yeast takes up oxygen from the medium. The point at which the maximum oxygen uptake was observed coincided with the maximum sterol concentration. Providing the oxygenation process was allowed to proceed until the maximum oxygen uptake rate was observed, or thereafter, the resultant yeast had no requirement for wort oxygenation to produce a standard fermentation profile. However, it was necessary to increase pitching rates by 30% in order to achieve the same vessel residence times as control fermentations.

Process plant has been described which was shown to be suitable for oxygenating yeast slurries at a scale sufficient for pitching up to 2000 hl of high gravity wort. It was demonstrated that yeast slurries containing up to 40% suspended solids could be oxygenated successfully, the process being completed within 4 h at a temperature of 20°C. Oxygenated yeast withstood the rigours of storage at elevated temperatures less well than untreated pitching yeast. However, at the low temperatures typically used for yeast handling in production scale brewing there were no significant differences in the ability of oxygenated and non-oxygenated yeast to withstand periods of storage. Consequently, it was concluded that yeast should be oxygenated immediately after cropping from fermenter and then stored if necessary. This would avoid the possible risks which might be associated with oxygenating stressed yeast.

Production scale (1500 hl) lager fermentations were demonstrably more consistent compared with conventional wort oxygenated controls. Trial beers were indistinguishable from controls on the basis of organoleptic criteria. There were no significant differences in the concentrations of total esters and total higher alcohols between trial and control beers. However, the range of concentrations was less in the trial beers.

There was greater consistency of total yeast growth in the trial fermentations compared with controls. Most strikingly, in trial fermentations there was a marked reduction in the extent of yeast growth compared with controls. Thus, trial yeast crops were reduced by nearly 50%. This resulted in significant gains in the efficiency of individual fermentations; however, should a brewery be totally devoted to yeast oxygenation the low growth rates could produce supply problems. Surprisingly, the reduced growth was not accompanied by changes in beer flavour.

The reasons for the low growth are obscure. The quantity of sterol synthesised during oxygenation was of the same magnitude as that seen in conventional wort oxygenated fermentations. However, the principal sterols formed during oxygenation were zymosterol and ergosterol. In fact, during oxygenation zymosterol increased as a proportion of the total sterol whereas ergosterol decreased. This could have some significance in that ergosterol is the principal free sterol in yeast, and as such is incorporated into membranes, whereas zymosterol is usually esterified[6,7]. It is possible, therefore, that less free sterol was available for direct incorporation into the membrane compared with a wort oxygenated fermentation. Nevertheless, this does not explain why fermentations which were provided with combinations of oxygenated yeast and oxygenated wort still grew less yeast than fermentations supplied with oxygen to the wort only. The superficial conclusion is that normal growth requires simultaneous exposure of yeast to wort and oxygen.

Acknowledgements

The authors thank the Directors of Bass Brewers for permission to publish this chapter. The contributions of Wendy Box, David Quain and Rory Bolton are gratefully acknowledged.

References

(1) Boulton, C.A. and Quain, D.E. (1987) Yeast, oxygen and the control of brewery fermentations. *Proceedings of the European Brewing Congress* **21**, 401–408.

(2) Quain, D.E. and Boulton, C.A. (1990) Oxygenated yeast. British Patent 2197341.

(3) Boulton, C.A., Jones, A.R. and Hinchliffe, E. (1991) Yeast physiological condition and fermentation performance. *Proceedings of the European Brewing Congress* **23**, 385–392.

(4) Quain, D.E., Thurston, P.A. and Tubb, R.S. (1981) The structural and storage carbohydrates of *S. cerevisiae*: changes during the fermentation of wort and a role for glycogen catabolism in lipid synthesis. *Journal of the Institute of Brewing* **87**, 108–111.

(5) Quain, D.E. and Haslam, J. (1979) The effects of catabolite derepression on the accumulation of steryl esters and the activity of 3-hydroxymethylglutaryl-CoA reductase in *S. cerevisiae*. *Journal of General Microbiology* **111**, 343–351.

(6) Weete, J.D. (1989) Structure and function of sterols in fungi. *Adv. Lipid Res.* **23**, 115–167.

(7) Rattray, J.B.M. (1988) Yeasts. In *Microbial Lipids*, Vol. 1, Ratledge, C. and Wilkinson, S.G. (eds), Academic Press, London, pp. 555–697.

(8) Quain, D.E. (1981) The determination of glycogen in yeast. *Journal of the Institute of Brewing* **95**, 315–323.

21 The Relationship Between Flocculation and Cell Surface Physical Properties in a *FLO1* Ale Yeast

MAUREEN RHYMES and KATHERINE SMART

Abstract The accumulation of lectin-like proteins on the cell surface of *FLO1* brewing yeast strains is required for flocculation to occur. The *FLO1* gene product (Flo 1p) comprises repeated sequences of threonine and serine which determine flocculence level. Both the C and N termini are hydrophobic; the former is believed to be the anchoring domain and the latter the active region containing the sugar recognition domain. Surface proteins are also reported to determine hydrophobicity in yeast, depending on the exposure of hydrophobic peptides. Although the mechanism for flocculation onset is not yet known, increased expression of this parameter has been reported to be a major determinant in flocculation. Examination of the ale strain NCYC 2593 using sugar and lectin inhibition failed to completely elucidate the flocculation phenotype; however, immuno-fluorescent staining with antibodies to Flo 1p revealed that this protein was indeed homogeneously expressed at the cell surface after 48 h growth. A haploid strain of *Saccharomyces cerevisiae* (V5) has been transformed to express the *FLO 1* gene containing part of and the entire repeated serine-threonine sequence. The relationship between the number of repeated sequences and extent of expression of cell surface characteristics and flocculation was examined. It was observed that flocculation ability and hydrophobic expression significantly correlated. Cell surface localization of Flo 1p and hydrophobic expression using immunofluorescent staining and hydrophobic bead attachment, respectively, revealed a close relationship between these parameters. No relationship with surface charge was apparent.

21.1 Introduction

Flocculation has been described as 'a series of complex phenomena influenced by diverse factors which are strain dependent and genetically controlled'[1]. Lectin-like interactions are suggested to be involved in the mechanism of flocculation and possibly controlled by the dominant *FLO 1* gene[2]. The flocculation protein is thought to bind to the side branches of α-mannan receptors of adjacent cells. This theory is supported by the fact that flocculation can be dispersed by sugars exerting competitive inhibition by binding to lectins[3]. Two entirely distinct mechanisms of lectin-like flocculation were initially proposed, involving a constitutive, mannose specific lectin in the *FLO1* type and a broad specificity lectin in the New *FLO* type, induced in stationary phase[3]. A third, mannose insensitive type has subsequently been described[4].

The *FLO 1* gene encodes for a large protein (Flo 1p), localized on the cell surface with a probable membrane-anchoring domain with a GPI attachment[1,5]. This protein also contains many repeated sequences of serine and threonine which are thought to determine flocculation intensity[1]. It was hypothesised that these sequences enabled the protein to span the cell wall and expose the N-terminal region to the surface. On the basis of amino acid sequence data this terminus is reported to be hydrophobic[6] and, as surface proteins are also reported to determine hydrophobicity[7], it has been suggested that flocculation ability may correspond to the degree of cell surface hydrophobic expression. In addition, increased expression of this parameter has been observed to be a major determinant in flocculation[8].

Immunofluourescent staining has previously demonstrated the deposition and localization of Flo 1 p in the cell wall of flocculent yeast strains[1,9] and, in order to verify the flocculation phenotype of the ale yeast NCYC 2593, this methodology was utilized in conjunction with genetically modified yeasts. These control strains were also used in new methodology to establish possible correlations between flocculation and cell surface physical properties.

21.2   Materials and methods

21.2.1   *Yeast strains and growth conditions*

The brewing yeast strain of *Saccharomyces cerevisiae* NCYC 2593 was obtained from the National Collection of Yeast Cultures (NCYC), Norwich. The haploid laboratory strains were supplied by Dr Bruno Blondin, Montpellier, France. The wild type strain V5 (*MAT* a, *ura3*) was transformed with yeast plasmids derived from YCp50 (*URA3*) carrying the *FLO1* gene and its own promoter (YEp*FLO1*). These transformed strains were denoted V5/p1, p4 and p13 and carried plasmids with 1, 4 and 13 copies of the A motif respectively[10].

Transformed V5 strains were grown in synthetic minimal medium which was uracil free to maintain the plasmids. Minimal media consisted of yeast nitrogen base without amino acids (YNB), with glucose 10%. The untransformed V5 strain (V5 p0) was grown in YNB as above with the addition of uracil, 50 mg/l. NCYC 2593 was grown in yeast extract peptone dextrose medium (YPD) or YNB (as untransformed strain) as required.

21.2.2   *Surface charge*

Cell surface charge was measured using the amine-modified latex bead assay, according to the method of Rhymes[11], involving the attachment of amine-modified latex microspheres to negatively charged cells. The relative population of charged cells, charged coupling capacity (CCC), was expressed as the percentage of colony forming units that had three or more spheres attached (100 colony forming units (cfu) evaluated).

21.2.3   *Surface hydrophobicity*

Cell surface hydrophobicity was measured using the hydrophobic latex microsphere attachment assay, a modification of the method of Hazen and Hazen[12]. The relative population of hydrophobic cells, hydrophobic coupling capacity (HCC) was expressed as the percentage of colony forming units that had three or more spheres attached[13] (100 cfu evaluated). Cell surface hydrophobicity was also assessed by the magnobead assay according to the method of Straver and Kijne[14].

21.2.4 *Flocculation*

Flocculation competence was assessed by a modification of the quantified Helm's flocculation test[15].

21.2.5 *Immunofluorescence staining*

Polyclonal anti-Flo 1p antibodies were raised against an *Escherichia coli* – expressed form of the protein by rabbit immunization as described by Bidard *et al.*[1] and supplied by Dr Bruno Blondin, Montpellier, France. Immunofluorescence staining followed the procedures described by Watzele *et al.*[16]

21.3   Results and discussion

21.3.1 *Examination of the relationship between* FLO 1 *gene modifications and cell surface physical characteristics*

The flocculation ability and level of expression of surface physical properties were determined on NCYC 2593 and genetically modified V5 strains after incubation for 48 h, at 25°C, in aerated conditions, in synthetic minimum media YNB.

In a typical *FLO 1* gene, 18 repeats of the serine-threonine sequence are present. The number of these repeated sequences influences flocculation potential (Fig. 21.1(a)). There was no significant difference between the flocculation ability of V5, p0 and p1 (11.74 $\pm$ 3.50% and 11.70 $\pm$ 2.87%, respectively), however, the ability of p4 was significantly (P <0.01) lower (6.58 $\pm$ 3.68%). The ability of both p13 and NCYC 2593 was significantly (P <0.01) higher, with p13 being the most flocculent (93.16 $\pm$ 1.11%). While NCYC 2593 was also highly flocculent, competence was significantly (P <0.01) lower (84.80 $\pm$ 3.15%) than that of p13. These observations supported the data of Bidard *et al.*[1]

Hydrophobic expression is shown in Fig. 21.1(b). The latex bead attachment assays correlated significantly (P <0.05) with flocculation ability (Pearson's correlation coefficient r = 0.99). The 3 strains (V5 p0, p1 and p4) which were the least flocculent also expressed low HCC (12.56 $\pm$ 5.32%, 3.29 $\pm$ 0.51% and 3.00 $\pm$ 1.00%, respectively). The highly flocculent strains p13 and NCYC 2593 exhibited significantly (P <0.01) higher expression, with slight but not significant differences in HCC (81.46 $\pm$ 7.49% and 74.38 $\pm$ 1.61%, respectively).

Surface charge expression is shown in Fig. 21.1(c). CCC was low for all the strains examined, but the capacity of V5 p0 (9.00 $\pm$ 2.65%) was significantly (P <0.05) higher than that of the other strains. No obvious relationship with hydrophobicity or flocculation was evident.

21.3.2 *Elucidation of the flocculation phenotype of the ale brewing yeast NCYC 2593*

The ale yeast NCYC 2593 is categorized as *FLO 1* genotype (NCYC, 1990); however, both sugar inhibition and lectin assays failed to substantiate this flocculation mechanism (results not shown). Immunofluorescent staining was therefore used to

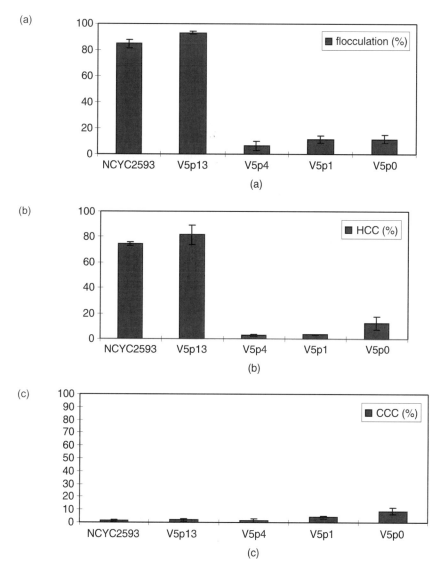

Fig. 21.1 Assessment of flocculation, hydrophobicity and charge. The level of expression of cell surface characteristics of NCYC 2593 and V5 control strains after growth for 48 h in YNB. Data are expressed as mean percentage (a) flocculation, (b) HCC and (c) CCC, with standard deviation.

verify the flocculation phenotype by demonstrating the presence and localization of cell surface Flo 1p.

21.3.2.1 *Fluorescent labelling with Flo 1p antibodies.* In order to confirm the efficacy of the antibodies, two of the V5 strains were selected as known flocculation controls; the untransformed V5 p0 as the non-flocculent control not expressing Flo 1p and p13 as the flocculent control expressing high levels of Flo 1p. Immunofluorescent staining was performed according to the method described by Watzele *et al.*[17] and cells were examined for surface labelling after growth in YNB for 48 h. Examination of immunofluorescent labelling during growth is shown in Plate 8. Very small, discrete areas of labelling were observed on only 20% of cells of the non-flocculent control (V5 p0) while a greater percentage of both p13 and NCYC 2593 were fluorescent (88.76% $\pm$ 7.91 and 91.0% $\pm$ 9.90, respectively) with 55% and 63% (respectively) of those cells homogeneously labelled.

These observations indicated that NCYC 2593 expressed Flo 1p. Primary deposition in flocculent strains was at the neck junction of mother and daughter, in emerging buds and at poles of single cells, followed by homogeneous distribution in stationary phase.

21.3.3 *Examination of the cell surface localization of hydrophobicity*

The localization of attachment of latex microspheres during the HCC assays was observed to correspond to similar areas as those in surface labelling during immunofluorescence assessment. Highly hydrophobic cells appeared to have randomly attached microspheres, heterogeneously scattered over the surface, whereas attachment of one or two beads was often observed at the poles or necks of mother and daughter cells.

Localization of hydrophobic expression was assessed on all three strains during growth, using hydrophobic bead attachment assays. Localization of hydrophobic beads closely corresponded to localization of Flo 1p. After incubation of NCYC 2593 for 4 h, attachment was randomly distributed, but polar attachment to single cells was evident. After 15 h incubation, a greater proportion of cells were observed to have homogeneous attachment, and localization was similar to Flo 1p labelling with attachment to necks of mother and daughter cells and polar attachment. After 48 h incubation beads attached randomly to most cells. Similar patterns of attachment were observed for V5 p13. Localization of bead attachment to the non-flocculent V5 p0 was difficult to assess as these cells were spherical.

21.3.4 *Correlation of presence of surface* Flo 1 *protein, flocculation and hydrophobic expression*

It was subsequently hypothesised that the extent of hydrophobic expression as well as flocculation level may be dependent on the quantity of Flo 1p exposed at the cell surface. Therefore these parameters were determined in NCYC 2593 and the control strains during exponential and stationary growth phases.

When comparing the magnitude of expression of all these parameters (Table 21.1), the number of cells with Flo 1p labelling correlated significantly ($P < 0.05$) with flocculation (Pearson's correlation coefficient $r = 0.99$, $P < 0.05$); and flocculation with HCC (Pearson's correlation coefficient $r = 0.99$, $P < 0.05$). Fluorescent labelling, flocculation ability and hydrophobic expression in NCYC 2593 and the flocculent control increased with growth. The HCC and magnobead assays also correlated significantly (Pearson's correlation coefficient $r = 0.999$, $P < 0.05$).

Table 21.1 Cell surface assessment of *Flo 1* protein, flocculation and hydrophobicity. Strains V5 untransformed (V5 p0), V5 transformed (V5 p13) and NCYC 2593 grown for 48 h in YNB at 25°C. Data are expressed as mean percentage with standard deviation. Partial labelling is expressed as a percentage of the total number of labelled cells.

Yeast strain	Growth time (h)	Total number of labelled cells (%)	Partially labelled cells (%)	Flocculation (%)	Hydrophobicity (HCC) (%)	Hydrophobicity (magnobead assay) (%)
V5 p0	15	21.00 ± 1.41	0.50 ± 0.71	7.46 ± 5.42	20.67 ± 6.59	23.58 ± 6.56
	48	20.00 ± 3.00	2.00 ± 1.00	11.32 ± 2.19	18.01 ± 8.03	7.56 ± 10.67
V5 p13	15	74.62 ± 9.46	25.23 ± 8.91	56.47 ± 3.25	71.20 ± 15.42	74.37 ± 6.35
	48	88.76 ± 7.91	49.81 ± 15.23	85.86 ± 2.92	88.40 ± 9.86	76.55 ± 3.75
NCYC 2593	15	77.75 ± 5.56	20.75 ± 6.50	54.07 ± 5.56	70.40 ± 6.80	62.26 ± 7.71
	48	91.00 ± 9.90	57.50 ± 9.19	84.40 ± 2.50	85.40 ± 6.58	76.75 ± 10.19

21.3.5 *Cell surface localization of flocculation protein and hydrophobicity*

The localization of Flo 1p and hydrophobicity on the surface of NCYC 2593 after growth for 48 h was also examined using confocal microscopy. Confocal micrographs revealed cells with immunofluorescent labelling and bead attachment in adjacent areas while unlabelled cells were observed to have negligible bead attachment.

21.3.6 *Assessment of the hydrophilic fraction*

In order to further investigate the relationship between these parameters, the hydrophobic and hydrophilic cell fractions were separated using the magnobead assay[14]. Flo 1 protein expression was much less evident in the hydrophilic fraction of both the flocculent strains, and a greater proportion of these cells were only partially labelled. Similar labelling in both hydrophilic and whole population was observed for the non-flocculent control strain.

21.4 Conclusions

The extent of expression of flocculation and hydrophobicity was closely related to the number of repeated gene sequences of serine and threonine in the genetically modified strains, supporting the observations of Bony *et al.*[9] The flocculation phenotype of NCYC 2593 was confirmed as Flo 1, and spacial distribution of Flo 1p during growth was observed to be similar to that previously reported[9,17]. Assessment of hydrophobic expression with latex beads revealed similar localization and distribution to cell surface Flo 1p, indicating the close relationship between these parameters. The degree of flocculation ability, hydrophobic expression and quantity of cell surface Flo 1p expression significantly correlated; however, no relationship with surface charge was observed. Examination of the hydrophilic cell population revealed a large reduction in homogeneously labelled cells, indicating that the surface protein responsible for flocculation ability is probably also responsible for hydrophobic expression. This hypothesis supports the observations that hydrophobicity is a major determinant of flocculation[8,18-20].

Acknowledgements

The authors would like to thank Sheona Bellis for help with photography and Janet Evins and Barry Martin for fluorescence microscopy assistance.

References

(1) Bidard, F., Bony, M., Blondin, B., Dequin, S. and Barre, P. (1995) The *Saccharomyces cerevisiae FLO 1* flocculation gene encodes for a cell surface protein. *Yeast* **11**, 809–822.

(2) Miki, B., Hung Poon, N., James, A. and Seligy, V. (1982a) Possible mechanism for flocculation interactions governed by gene *FLO1* in *Saccharomyces cerevisiae*. *Journal of Bacteriology* **150**, 878–889.

(3) Stratford, M. (1989) Evidence for two mechanisms of flocculation in *Saccharomyces cerevisiae*. In *Seventh International Symposium on Yeasts*, pp. S441–445.

(4) Masy, C.L., Henquinet, A. and Mestdagh, M.M. (1992b). Flocculation of *Saccharomyces cerevisiae*: inhibition by sugars. *Canadian Journal of Microbiology* **38**, 1298–1306.

(5) Bony, M., Thines-Sempoux, D., Barre, P. and Blondin, B. (1997) Localization and cell surface anchoring of the *Saccharomyces cerevisiae* flocculation protein Flop. *Journal of Bacteriology* **179**, 4929–4936.

(6) Watari, J., Takata, Y., Ogawa, M., Sahara, H., Koshino, S., Onnela, M., Airaksinen, U., Jaatinen, R., Pentilla, M. and Keranen, S. (1994) Molecular cloning and analysis of the yeast flocculation gene *FLO1*. *Yeast* **10**, 211–225.

(7) Hazen, K.C. and Hazen, B.W. (1993) Surface hydrophobic and hydrophilic protein alterations in *Candida albicans*. *FEMS Microbiology Letters* **107**, 83–88.

(8) Straver, M.H., van der Aar, P.C., Smit, G. and Kijne, J.W. (1993) Determinants of flocculence of brewer's yeast during fermentation in wort. *Yeast* **9**, 527–532.

(9) Bony, M., Barre, P. and Blondin, B. (1998) Distribution of the flocculation protein, Flop, at the cell surface during yeast growth: the availability of Flop determines the flocculation level. *Yeast* **14**, 25–35.

(10) Bidard, F., Blondin, B., Dequin, F., Vezinhet, F. and Barre, P. (1994) Cloning and analysis of a *FLO5* flocculation gene from *Saccharomyces cerevisiae*. *Current Genetics* **25**, 196–201.

(11) Rhymes, M.R. (1999) The effect of starvation on brewing yeast cell surface physical characteristics. M.Phill. thesis, Oxford Brookes University.

(12) Hazen, K.C. and Hazen, B.W. (1987) A polystyrene microsphere assay for detecting surface hydrophobicity variations within *Candida albicans* populations. *Journal of Microbiological Methods* **6**, 289–299.

(13) Smart, K.A., Boulton, C.A., Hinchcliffe, E. and Molzahn, S. (1995) Effect of physiological stress on the surface properties of brewing yeasts. *Journal of the American Society of Brewing Chemists* **53**, 33–38.

(14) Straver, M.H. and Kijne, J.W. (1996) A rapid and selective assay for measuring cell surface hydrophobicity of brewer's yeast cells. *Yeast* **12**, 207–213.

(15) Bendiak, D.S. (1994) Quantification of the Helm's flocculation test. *Journal of the American Society of Brewing Chemists* **52**, 120–122.

(16) Watzele, M., Klis, F. and Tanner, W. (1988) Purification and characterization of the inducible *a* agglutinin of *Saccharomyces cerevisiae*. *The EMBO Journal* **71**, 1483–1488.

(17) Masy, C.L., Henquinet, A. and Mestdagh, M.M. (1992) Fluorescence study of lectin-like receptors involved in the flocculation of the yeast *Saccharomyces cerevisiae*. *Canadian Journal of Microbiology* **38**, 405–409.

(18) Mestdagh, M., Rouxhet, P. and Dufour, J. (1990) Surface chemistry and flocculation of brewing yeast. *Ferment* **3**, 31–37.

(19) Smit, G., Straver, M., Lugtenberg, B. and Kijne, J. (1992) Flocculence of *S. cerevisiae* cells is induced

by nutrient limitation, with cell surface hydrophobicity as a major determinant. *Applied and Environmental Microbiology* **58**, 3709–3714.

(20) Azeredo, J., Ramos, I., Olivera, R. and Teixeira, J. (1997) Yeast flocculation: a new method for characterising cell surface interactions. *Journal of the Institute of Brewing* **103**, 359–361.

22 Physiological Changes in the Yeast Cell Wall and Membrane Structure and Function During Yeast Storage and From the Fermentation Process

ALDO LENTINI and PETER ROGERS

Abstract The physiological status of a brewing yeast cell, in particular its cell wall and structure and function, may be significantly influenced by its environment if it is exposed to stress during the various process stages of beer production. This in turn can have a significant impact on the overall vitality and viability of the yeast, and subsequently the overall physical and flavour quality of the final beer product. By understanding the physiological changes that occur within the yeast, improvements in yeast quality (health and activity) can be achieved.

The current study examines the yeast cell and changes that occur within the cell membrane (sterol and fatty acid composition) and cell wall (e.g. carbohydrate, protein and surface characteristics) when the yeast is exposed to various adverse conditions during yeast storage and fermentation. The various stress effects on the yeast can be caused by higher temperatures, high ethanol levels, variable slurry pH and limited nutrient availability over a specified time period.

The main part of the investigation was to note the effects these stress factors had on the cell membrane and the relationship between membrane composition and fluidity (i.e. the levels of sterols, saturated fatty and unsaturated fatty acids), and its subsequent impact on sugar uptake, nutrient utilisation, protease release and overall viability and vitality. This study has shown that yeast increases membrane fluidity in order to protect its health status when stressed during storage and fermentation.

This information is being used to optimise the overall environmental conditions for each specific brewing yeast strain used at Carlton and United Breweries. By understanding the relationship between cell wall/membrane structure and specific biological functions related to sugar and nutrient uptake metabolism, improvements in process conditions can be designed to improve yeast viability and vitality. This will also improve the overall performance of our yeasts in the brewery. This has already resulted in greater consistency in fermentation performance and improved yeast usage within our breweries.

23 Optimisation of Storage and Propagation for Consistent Lager Fermentations

GAVIN HULSE, GEORGIA BIHL, GONTSE MORAKILE and
BARRY AXCELL

Abstract Storage of master yeast cultures in liquid nitrogen and working cultures at $-70°C$ has been shown to be an attractive alternative to traditional methods (such as slopes) of storing working yeast cultures. It was noted that the above storage conditions selected against respiratory deficient culture variants. By using standardised working cultures as the inoculum for the first stage of laboratory propagation it was shown that the growth and performance in the subsequent stages was more consistent. This consistency allowed for the predictability of plant propagation, with regard to top-up and transfer times. Fermentations from yeast stored and cultured in the above manner were shown to be more consistent than those fermentations from slope derived cultures.

23.1 Introduction

The development of an industrial microbiological process requires that consideration be given to the preservation and maintenance of the production strain(s) being employed[1]. Indeed the aims of yeast propagation can best be summarised by the following statement from Voight and Walla[2]: 'The prime objective of yeast propagation is to produce sufficient yeast in the shortest possible time of the desired physiological condition to adequately pitch the zero generation fermentation.' The yeast produced should be consistent, in terms of biomass produced and physiological condition from propagation to propagation. This will result in more consistent and predictable fermentations, leading to beer of a more consistent quality as demanded by the end consumer; this statement applies equally to the maintenance of yeast cultures. Deterioration of the performance of the production culture can result in considerable disruption to the production process, and most certainly financial losses will be incurred.

Many brewing companies have gone to extreme trouble to select the variant of a particular strain that performs to their desires in terms of matching the existing plant facilities and producing product that meets the customers' expectations in terms of flavour and taste. Therefore it is essential that culture storage methods should be reliable. The factors to be considered when selecting a preservation system include intrinsic microbiological factors and extrinsic factors such as the available expertise and equipment.

Historically many methods have been used to preserve cultures, including[1]: (1) subculturing, (2) desiccation, (3) lyophilisaion, and (4) cryopreservation. Historically we have used subculturing techniques as the basis for culture preservation. However, in our quest to improve our fermentation consistency, as described previously[3], yeast culture storage and propagation procedures were identified as leverage areas to improve fermentation consistency. Research[1,4] has indicated that yeast strains respond differently to both desiccation and lyophilisation. Cryopreservation appears to be the preferred method of yeast culture preservation[5].

Storage of yeast cultures, at each of the regional breweries, in liquid nitrogen ($-196°C$) was deemed to be impractical. With the availability of commercial freezers suitable for the storage of cultures at $-70°C$ it was proposed that the master cultures be stored at $-196°C$ and the regional breweries store a suitable supply of quality assured working cultures at $-70°C$. This required determining the suitability of storing our particular yeast strain at both $-196°C$ and at $-70°C$.

From the above it was concluded that there was a need to develop a master culture and a working culture storage system (Fig. 23.1).

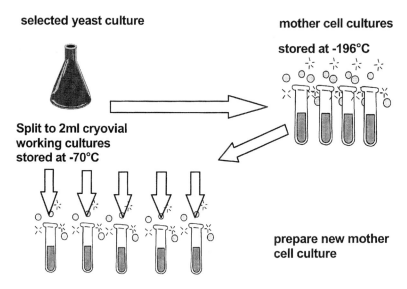

Fig. 23.1   Schematic representation of proposed yeast culture storage protocol.

23.2   Materials and methods

23.2.1   *Cryopreservation*

A suitable number of cryovials, containing 1.15 ml of a yeast suspension, were prepared for cryopreservation[6] to allow for frequent thawing and assessment of survival, variants and growth rates over the duration of a year.

23.2.2   *Growth curves*

Growth curves were determined using standard techniques. Production maltose adjunct wort was inoculated with a bottom fermenting production strain of *Saccharomyces cerevisiae*. Samples were taken 2 hourly and monitored for yeast counts using a standard haemocytometer counting chamber.

23.2.3 *Determination of optimum inoculum transfer times*

For each stage of laboratory propagation growth curves were determined. The time to reach the late logarithmic stage of growth was recorded. This information was used to determine the optimum transfer time for each stage of laboratory propagation.

23.2.4 *EBC tall-tube fermentations*

To determine the fermentation performance of the cryopreserved derived cultures as opposed to the slope derived cultures tall-tube fermentations were carried out using standard conditions.

23.2.5 *Production scale fermentations*

3000 hl production scale fermentations were carried out with yeast produced using the new optimised procedures. The fermentations were treated no differently from the standard production fermentations.

23.3 Results and discussion

23.3.1 *Cryopreservation*

The results of this work indicated that there was no selection for variants, particularly petites (Table 23.1). Indeed it appears that the petites do not survive long term cryopreservation as well as the parent variant. The reasons for this are at present not understood.

Table 23.1 Survival of variants and petites with time.

Before cryopreservation	After cryopreservation	
	30 h	1 year
Culture variants		
3.9%	3.6%	3.3%
6.7%	3.6%	4.0%
Petite mutants		
1.1%	1.4%	1.1%
1.9%	0	0.8%

23.3.2 *Growth curves*

Survival rates and growth curves were deemed to be acceptable (Fig. 23.2).

23.3.3 *Determination of optimum inoculum transfer times*

Figure 23.4 shows the growth curves obtained from the first stage (15 ml stage) of laboratory yeast propagation. The results show that there was little difference in the

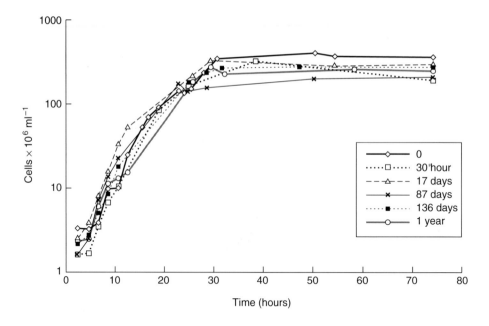

Fig. 23.2   Growth curves of cryopreserved cultures thawed at differing time intervals ($-196°C$).

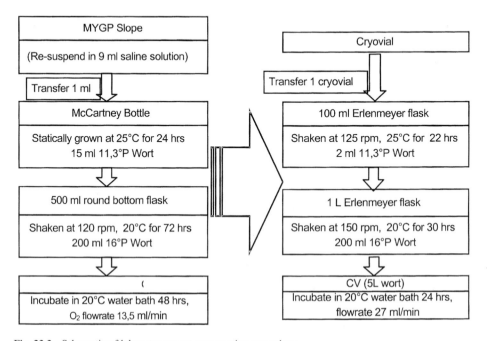

Fig. 23.3   Schematic of laboratory yeast propagation procedure.

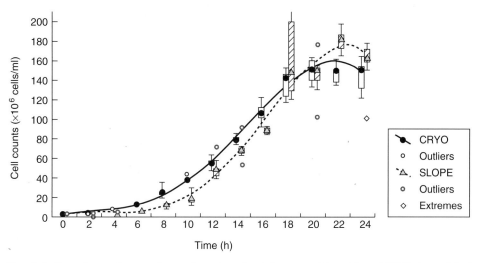

Fig. 23.4 Stage 1 growth curves from cultures inoculated from cryovials and from MYGP slopes (15 ml stage grown in 100 ml Erlenmeyer flask and shaken at 150 min^{-1}.

growth of the yeast from the two storage systems (cryopreserved in liquid nitrogen versus storage at 4°C on MYGP slopes). The cryopreserved cultures exhibited a shorter lag phase but did not reach the same final cell density. The consistency of growth (average of 10 cultures for each curve) as indicated by the error bars was vastly improved in the cryopreserved cultures as opposed to the cultures derived from the subcultured slopes. In the instance of the slopes all ten growth curves were derived from ten slopes and the curves from the cryopreserved cultures were from ten individual

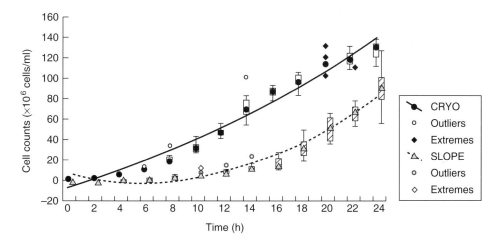

Fig. 23.5 Stage 2 growth curves from cultures inoculated from cryovials and from MYGP slopes (pre-culture: 15 ml 11.3°Plato wort in 100 ml Erlenmeyer flask and transferred into 200 ml 16°Plato wort).

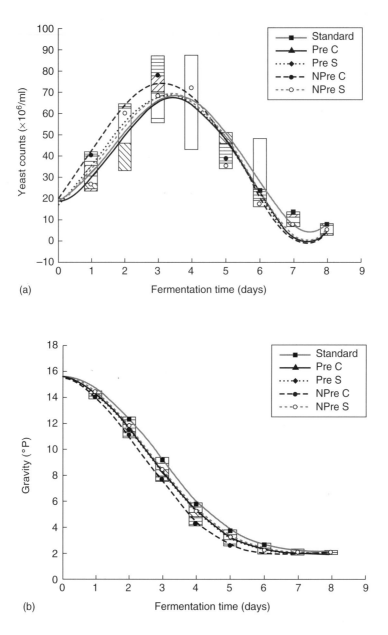

Fig. 23.6 Attenuation curves of zero generation EBC tall-tube fermentations: (a) yeast count and (b) gravity.

cryovials. The prime objective of the study was to improve the consistency between subsequent propagations (biomass yield and physiological status of the propagated yeast). From this stage it was apparent that the cultures derived from the cryopreserved cultures were more consistent than those derived from the MYGP slopes.

From Fig. 23.5 it is clear that the performance of the cryopreserved cultures was far superior when compared with the MYGP slope derived cultures. Currently we cannot offer any explanations for this observation, but we have observed it repeatedly. In addition to the improved growth performance the degree of consistency of perfor-

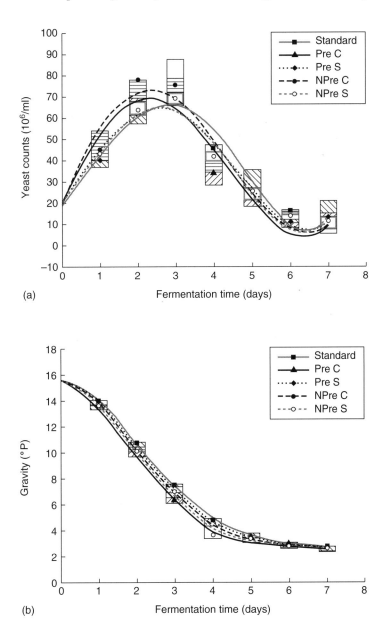

Fig. 23.7 Attenuation curves of first generation EBC tall-tube fermentations: (a) yeast count and (b) gravity.

mance was superior (as indicated by the error bars in Fig. 23.5) in the cryopreserved derived cultures.

From Fig. 23.6 it is apparent that the cryopreserved cultures outperformed (improved consistency of fermentation patterns and improved attenuation) the MYGP slope derived cultures. This benefit was observed on repitching the cropped zero generation yeast into first generation fermentations (Fig. 23.7).

23.4   Conclusions

In the above study it was shown that by optimising both culture storage and yeast propagation the resultant fermentation performance and reproducibility of fermentations were improved. The provision of working cultures that are identical, in terms of initial starting biomass and physiology, allows one to ensure that all subsequent transfers occur at the optimum late logarithmic stage of growth. It is essential that transfers occur before the onset of any limitations that may induce stress responses that will be carried forward to the next stage of propagation[3]. These stress responses will adversely affect the performance of the yeast in the subsequent stage of propagation. This was evidenced by lack of reproducibility, extended lag times, slower growth and less biomass produced. During the study the potential for decreasing the number of transfers by eliminating the first stage (15 ml stage) was investigated. To date the results have indicated that the direct transfer into high gravity wort with the subsequent larger dilutions ($100 \times$) did not adversely affect the growth responses. Indeed initial work has shown that the same total biomass can be obtained in a shorter time period by eliminating the first stage of laboratory yeast propagation.

By knowing the initial starting biomass and the growth rate one can apply the growth equation in order to determine the optimum starting biomass concentrations to ensure that the stages are ready for transfer at convenient working times. It also allows one to adjust the dilutions dependent upon in-process delays. This offers significant advantages in improved reproducibility of production scale fermentations.

The production scale fermentations showed improved reproducibility of fermentation performance, improved attenuation levels and improved yeast quality at cropping (data not presented). Currently this investigation has allowed us to modify and improve our laboratory yeast propagation procedures to that represented in Fig. 23.3.

Acknowledgements

We are grateful to the Directors of South African Breweries for permission to publish this work, to the University of the Orange Free State for the initial cryopreservation work, and to Prospecton Brewery for carrying out the production scale evaluations.

References

(1) Kirsop, B.E. and Doyle, A. (1991) *Maintenance Microorganisms and Cultured Cells. A Manual of Laboratory Methods*, Academic Press, London.

(2) Voight, J.C. and Walla, G. (1995) A novel yeast propagation system. *Proc. Inst. Brew.* (5th C & SA Sect., Victoria Falls), 173–178.

(3) Hulse, G.A., Quilliam, W. and Cosser, K. (1997) Practical observations on yeast propagation transfer times and resultant fermentation performance. Presented at the 1st Brewing Yeast Fermentation Performance Congress, Oxford Brookes University, Oxford.

(4) Russell, I. and Stewart, G.G. (1981) Liquid nitrogen storage of yeast cultures compared to more traditional storage methods. *Journal of the American Society of Brewing Chemists* **39**(1), 19–24.

(5) Walker, G.M. (1998) *Yeast Physiology and Biotechnology*, Wiley, Chichester.

(6) Kock, J.L.F., Morakile, G.I., Pretorius, E.E. and Pohl, C.H. (2000) Maintenance of brewing inocula through cryopreservation.

24 The Effect of Wort Maltose Content on Volatile Production and Fermentation Performance in Brewing Yeast

OMAR YOUNIS and GRAHAM STEWART

Abstract   Previous work has established that the metabolism of maltose produces fewer volatiles compared with levels obtained from the fermentation of glucose or fructose when yeast is fermented in synthetic media in shaking flasks, and several hypotheses for this are considered. Observed differences in yeast viability and vitality, and possible reasons for this, are discussed.

The use of very high maltose (VHM) adjuncts, in the form of syrups, and its effects were investigated. Syrup was added at levels of between 30% and 35% containing up to 70% maltose to high gravity (20°Plato) worts. It was found that static fermentations in glass tall-tubes also resulted in a reduced level of volatile production, compared with similar gravity all-malt brews. After dilution of the high gravity adjunct ale beer to the same ethanol concentration as that obtained from sales gravity brews, there remained an increased level of several esters in the 30% VHM adjunct beer, while higher alcohol production in many cases fell to below the level obtained from the all-malt sales gravity brew. This phenomenon was strain dependent. Addition of nutrients to the wort in order to achieve higher alcohol production is discussed.

Several two-hectolitre fermentations were conducted in order to assess whether the observed differences in volatile production were applicable at the pilot brewery scale. It was found that for both 12°Plato and 20°Plato brews there were fewer volatiles formed from fermentation of 30% VHM syrup adjunct wort compared with all-malt wort. To determine whether or not these differences were detectable to the consumer, a taste panel was assembled which compared the six beers in a series of triangular taste tests. Results indicated that the differences observed at the analytical level could be detected by the panellists, particularly when the 20°Plato brewed beers were evaluated.

24.1   Introduction

The use of high gravity (16 to 20°Plato) brewing has been implemented in many breweries worldwide and its relative merits and drawbacks have been documented[1]. One such disadvantage of the process is the difficulty in obtaining an acceptable flavour match to existing products produced using normal gravity (10 to 14°Plato) wort due to the disproportional increase in volatile production encountered in high gravity brewing. Therefore it is often necessary to make adjustments to the brewing process (e.g. fermentation temperature, yeast pitching rate, wort carbohydrate profile, etc.) in order to overcome such inconsistencies. However, as volatile production by yeast is a strain dependent phenomenon, it is not always possible to obtain an acceptable flavour match by confining the adjustments to the aforementioned parameters.

This study examined the production of volatile compounds by different strains of brewing yeast in synthetic media, all-malt brewer's wort and brewer's wort containing varying degrees of very high maltose (VHM) syrup (which typically contained 70% maltose) as adjunct. In order to obtain an improved flavour match to normal gravity brewed beer it was necessary to supplement the adjunct wort with various metals and amino acids. Yeast strains studied exhibited differing fermentation performance depending on the media being employed. Lower levels of volatiles were produced when VHM syrup adjunct wort was fermented compared with fermentation of glucose syrup adjunct wort or all-malt wort. A trained taste panel was established to determine whether or not these differences could be detected by the consumer.

24.2 Materials and methods

24.2.1 Yeast maintenance

Brewing ale and lager yeast strains of *Saccharomyces cerevisiae* were employed in this study. All strains were maintained on 10% glucose-peptone-yeast extract (PYN) media agar slopes, stored at 4°C and subcultured every 3 to 4 months.

24.2.2 Fermentation media and conditions

For peptone-yeast extract-nitrogen (PYN) fermentations, conditions were as previously reported[2]. In 1.5 litre brewer's wort fermentations using tall-tubes, yeast was pitched at 3.5 g/l and 6.5 g/l for 12°Plato and 20°Plato wort respectively[3]. Stock solutions of Ca^{2+}, Mg^{2+}, Zn^{2+} and Fe^{2+} were prepared and added to wort before autoclaving to give final concentrations of between 1 and 50 ppm. Amino acids (valine, leucine and isoleucine) were added to sterile wort via sterile filtration to give final concentrations of added amino acids of 0.2, 0.5, 0.75, 1.0 and 5.0 mM. For pilot brewery fermentations, a pre-brew was conducted in order to obtain sufficient biomass. Yeast was pitched at similar levels to tall-tube fermentations. Six 2 hl brews were conducted: 12°Plato all-malt, 12°Plato (30% syrup adjunct), 12°Plato (30% VHM syrup adjunct), 20°Plato all-malt, 20°Plato (30% syrup adjunct) and 20°Plato (30% VHM syrup adjunct).

24.2.3 Volatile production

Esters and higher alcohol production was followed by headspace gas chromatography[2].

24.2.4 Viability and vitality of yeast

Yeast cell viability was measured using alkaline methylene blue staining. Results were confirmed by conducting colony counts on agar plates and also by employing methylene violet (3-RAX) staining. Cell vitality was determined using the modified acidification power test[2].

24.2.5 Glycogen and trehalose levels

The method employed to determine the concentration of the storage carbohydrates glycogen and trehalose was that of Parrou and François[4]. Carbohydrate was degraded to glucose residues, the concentration of which was determined using glucose oxidase (Sigma) and read at 420 nm using an ELISA plate reader.

24.2.6   *Acetyl-CoA levels*

Acetyl-CoA levels were determined using the method devised by Thurston and Quain[5]. Fluorescence was measured using a Jenway Model 6200 Spectrofluorimeter.

24.2.7   *Taste panel*

A trained taste panel consisting of approximately 20 members was assembled. The six pilot brewery brewed beers were diluted to similar ethanol levels and compared in a series of triangular taste tests[6]. Additionally, beers were assessed for individual flavour notes, e.g. estery character, sweetness, sulfur notes, by several members of the taste panel.

24.3   Results and discussion

24.3.1   *Synthetic media fermentations*

Several strains of *S. cerevisiae* (ale and lager type) were fermented in shaking flasks in PYN medium containing 4% glucose, fructose or maltose as the sole carbohydrate source. As can be seen from Fig. 24.1, metabolism of maltose produced lower levels of ethyl acetate, with no significant differences in production obtained from fermentation with glucose or fructose. A similar pattern was observed with all other volatiles studied, while ethanol levels did not vary considerably regardless of the carbohydrate source. The reasons why fewer volatiles are formed from the metabolism of maltose are unclear but there are several possibilities. The cells take up similar levels of sugar, although at differing rates, with maltose uptake being slower. If different levels of glycolytic intermediates are formed upon utilisation of the two sugars, then this could lead to the formation of lower levels of acetyl-CoA, and thus less substrate for ester synthesis. Figure 24.2 shows that cells grown in maltose PYN do contain lower levels

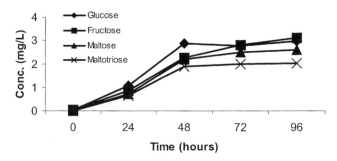

Fig. 24.1   Production of ethyl acetate by *S. cerevisiae* (ale type) when a synthetic medium containing glucose, fructose maltose or maltotriose was employed as the sole carbohydrate source.

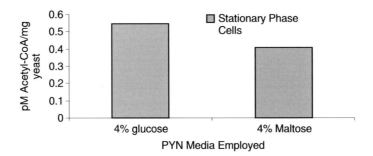

Fig. 24.2 Relative levels of acetyl-CoA obtained from yeast cells grown in 4% glucose or 4% maltose.

of acetyl-CoA than similar cells grown in glucose PYN. However, although this may explain why fewer esters are formed, the lower levels of higher alcohol production by maltose grown cells would still be unexplained.

If cells grown in different media had an effect on the intracellular pH, then it is possible that this could effect various cellular enzyme activities, including those involved in ester synthesis. Figure 24.3 shows the intracellular pH of yeast cells grown in 4% glucose and 4% maltose. Although differences in intracellular pH were observed, these could be explained by the differences in osmotic pressure exerted on the cells by the different media, as the glucose medium will contain twice as many molecules as maltose PYN and thus exert twice the osmotic pressure. If this were the case, then fermentation with PYN containing 4% maltotriose should result in the production of fewer volatiles than with maltose fermentation (Fig. 24.1).

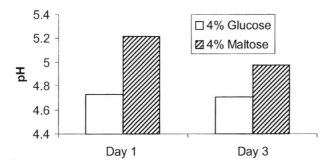

Fig. 24.3 Intracellular pH of yeast cells grown in PYN containing 4% glucose or 4% maltose.

The fermentation performance of the yeast cells also varied depending on which medium was being fermented. Cells grown in glucose (2% and 4%) containing PYN consistently had lower viabilites and vitalities than those grown in similar concentrations of maltose containing PYN. It was thought that this may be due to increased levels of storage carbohydrate being formed by cells which are growing in maltose containing media, as had been previously reported[7]. This was confirmed in this study with cells grown in maltose containing higher levels of trehalose than those grown in glucose (Fig. 24.2). Intracellular glycogen levels did not vary significantly.

24.3.2   Brewer's wort fermentations

Results obtained from fermentations using synthetic media were also applicable to wort. Wort containing higher levels of maltose (from the use of very high maltose syrup adjunct) resulted in fewer volatiles being formed and the yeast cells having higher viabilities and vitalities compared with cells grown in all-malt wort.

One of the aims of this study was to design a high gravity wort which would give beer (following appropriate dilution) a closer flavour match to all-malt 12°Plato wort. In order to achieve this, the wort fermentation temperature was altered, as was the wort amino acid and metal ion composition.

Addition of zinc greatly increased the rate of fermentation and the production of all volatiles. Although the use of supplements with defined amino acid profiles is questionable on a production basis at present, such supplements are employed in other fermentation industries (e.g. antibiotics). In the future it is anticipated that this cost will reduce, making it economically viable for use in brewing. It was possible to increase the production of specific higher alcohols and the corresponding ester by supplementing wort with certain amino acids (Fig. 24.4).

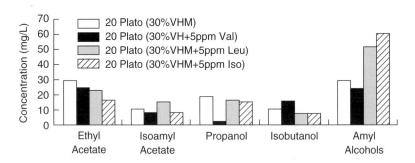

Fig. 24.4   Effect of addition of specific amino acids on volatile production in *S. cerevisiae* (ale type).

24.3.3   High gravity wort design for improved flavour matching

Previous work in this laboratory[3] has shown that fermentation of 20°Plato wort, which contained 30% VHM syrup as adjunct, produced fewer esters than all-malt 20°Plato wort but more than all-malt 12°Plato (after dilution to equivalent ethanol levels). Therefore in order to obtain an overall improved flavour match to all-malt 12°Plato beer, it was necessary to further reduce ester production. However, reduction of higher alcohol levels was not desired as the production of these volatiles was similar in all-malt 12°Plato wort and 20°Plato (30% VHM adjunct) wort. Fermentation of worts containing upwards of 30% adjunct was very slow and consequently not of use to the brewer. The addition of zinc was also considered unsuitable as its effects were non-specific, i.e. production of all volatiles studied was enhanced. It was judged that addition of 0.75 mM leucine, valine and 5 ppm Ca^{2+} to 30% VHM syrup adjunct wort would result in production of a beer with a more similar volatile profile

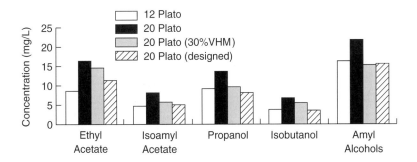

Fig. 24.5 Comparison of volatile production yeast fermenting all-malt 12°Plato, all-malt 20°Plato, 20°Plato (30% VHM syrup) and supplemented 20°Plato worts.

to all-malt 12°Plato beer than currently observed with all-malt 20°Plato wort. Figure 24.5 shows that there is an improvement in flavour matching when the supplemented adjunct wort is fermented.

24.3.4 *Pilot brewery fermentations and taste panel results*

The six beers which were brewed in the Centre's pilot brewery were compared by a 20 member trained taste panel over a period of several weeks (Table 24.1). The panel could not detect any significant differences between all-malt and 30% glucose syrup adjunct beers, when comparing 12°Plato and also 20°Plato brews. However, when comparing all-malt beer with 30% VHM syrup adjunct beer, there were significant differences detected by the panel in the case of sales and high gravity beers. This indicates that the differences in volatile production observed in small scale fermentations are also applicable at the 2 hL level and are significant enough to be detected by the consumer. Consequently, modifying wort maltose concentration could be used to control excessive volatile production, for example in high gravity brewing.

Table 24.1 Comparison of pilot brewery brewed beers using triangular taste test methodology.

Beer pairings	Probability of taste difference
All-malt 12°Plato versus 30% glucose syrup 20°Plato	No differences detected by taste panel
All-malt 20°Plato versus 30% glucose syrup 20°Plato	No differences detected by taste panel
All-malt 12°Plato versus 30% VHM syrup 12°Plato	95% probability that beers are different
All-malt 20°Plato versus 30% VHM syrup 20°Plato	95% probability that beers are different
30% VHM syrup 20°Plato versus 30% syrup 20°Plato	95% probability that beers are different
30% VHM syrup 20°Plato versus 30% syrup 12°Plato	95% probability that beers are different

24.4 Conclusions

The production of volatile compounds was reduced when maltose was fermented in a synthetic medium compared with metabolism of glucose or fructose, and also when the maltose:glucose ratio was increased by the addition of syrup adjunct. Fermentation in this elevated maltose medium results in improved yeast viability and vitality. The reasons for this are unclear, although it may simply be due to a decrease in the osmotic pressure which is exerted on the cells when maltose is fermented compared with hexose sugars.

Differences observed in volatile production could be detected by the consumer, highlighting the possibility of limiting ester levels (e.g. in high gravity brewing) if desired. It was possible to improve the flavour match between all-malt 12°Plato wort brewed beer and 20°Plato beer by modifying the wort carbohydrate, amino acid and metal ion spectra, although each such supplemented wort was specific to a particular strain.

Acknowledgements

The authors wish to thank the ICBD for financial assistance in carrying out this study, Scottish Courage Brewing Ltd., Edinburgh, for the supply of very-high maltose syrup and Dr. David Quain, Bass Brewers, for supplying the method for acetyl-CoA measurement. Technical assistance provided by Graham McKernan of the ICBD is greatly appreciated.

References

(1) Stewart, G.G. and Russell, I. (1998) In *An Introduction to Brewing Science and Technology*, Series III, *Brewer's Yeast*, The Institute of Brewing, London.
(2) Younis, O.S. and Stewart, G.G. (1998) Sugar uptake and subsequent ester and alcohol production in *Saccharomyces cerevisiae. Journal of the Institute of Brewing* **104**, 255–264.
(3) Younis, O.S. and Stewart, G.G. (1999) The effect of malt wort, very high gravity malt wort and very high gravity adjunct wort on volatile production in *Saccharomyces cerevisiae. Journal of the American Society of Brewing Chemists* **52**, 39–45.
(4) Parrou, J.L. and François, J. (1997) A simplified procedure for a rapid and reliable assay of both glycogen and trehalose in whole yeast cells. *Analytical Biochemistry* **248**, 186–188.
(5) Thurston, P.A. (1982) Lipid metabolism of *Saccharomyces cerevisiae* and control of volatile ester synthesis in wort fermentations. Ph.D. thesis.
(6) Analytica-EBC: Method 13.7, Sensory Analysis: Triangular Test.
(7) Panek, A.D., Sampaio, A.L., Braz, G.C. and Mattoon, J.R. (1979) Genetic and metabolic control of trehalose and glycogen synthesis. New relationships between energy reserves, catabolite repression and maltose utilization. *Cellular and Molecular Biology* **25**, 334–354.

25 Fermentation Modelling in Commercial Beer Manufacture, Including Temperature Effects

DAVID YUEN and PAUL AUSTIN

Abstract   The research described is based on a project undertaken at Lion Nathan's Auckland brewery, jointly by University of Auckland and Lion Nathan personnel. According to Lion's historical fermentation data records, the average fermentation length for beer A, the subject of this study, was 8.0 $\pm$ 1.2 days. Although the quality of the historical data available limited the extent of the analysis that could be undertaken, it was possible to identify the influence on fermentation length of different yeast propagation and generation number. To facilitate the design of a better control system to compensate for the effects of these and other variations, a fermentation model was established and validated. The model is comprised of two parts. In the first part, the sugar, yeast and temperature profiles are modelled through a series of differential equations. In the second part, the diacetyl concentration is predicted using a neural network model. The model describes the temperature dependence of the components particularly well: actual measured production data for beer B, which is fermented at a much higher temperature than beer A, is predicted accurately using the model developed from beer A data.

25.1   Introduction

The production of any brewery must be able to meet the continually varying market demand for beer. Currently, the primary fermentation time for a mainstream Lion Brewery product, beer A (see process flow diagram, Fig. 25.1), varies between 6.5 days and 9.5 days in 80% of all cases (Fig. 25.2). However, on rare occasions, the fermentation time can be as long as 12 days. A preliminary study of the historical data records for beer A revealed that the length of the fermentation was strongly linked with the propagation and generation of the pitching yeast. The uncertainty associated with the fermentation time makes production planning a difficult task and since the exact timing of beer supply cannot be guaranteed, extra stock must be held as a provision for the worst case batch length scenario. Furthermore, at times when demand for beer is high, the company has a particular interest in ensuring batch times are kept to a minimum. Therefore, it is very desirable, from the company's point of view, to reduce variations in fermentation length.

Attempts are being made to improve the yeast management scheme as a means of reducing the variations in fermentation length. However, this is not an easy task. The research reported here represents an attempt to shorten the variation in batch time by refining the general fermentation control strategy. A fermentation model has been developed as a necessary step in designing an improved control strategy.

malt

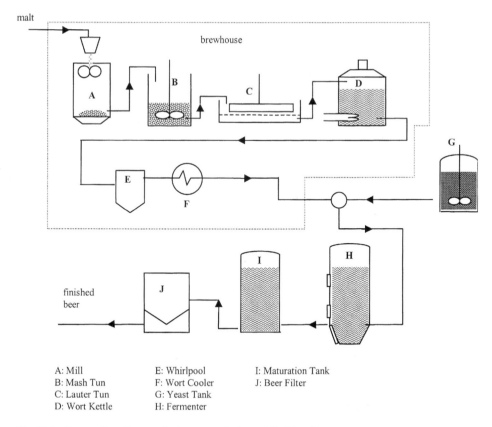

A: Mill   E: Whirlpool   I: Maturation Tank
B: Mash Tun   F: Wort Cooler   J: Beer Filter
C: Lauter Tun   G: Yeast Tank
D: Wort Kettle   H: Fermenter

Fig. 25.1   Process flow diagram for beer manufacture at the Lion Brewery.

25.2   Fermentation modelling

Beer is complex natural product, in which a large number of carbohydrates, alcohols, esters, organic acids and other flavour compounds are dissolved. Clearly, it is not possible to include all these variables in the model.

In this work, the concentrations of the five fermentable sugars (glucose, fructose, sucrose, maltose and maltotriose) and of the yeast content were the subjects of a differential equation-based fermentation model. Other important variables, such as real extract, specific gravity, apparent fermentation and ethanol content, are then able to be estimated from the modelled sugar concentration and yeast content, using appropriate correlations. Near the end of the beer A fermentation, the brewer usually has to wait for the diacetyl decomposition to fall below a minimum desired value. Therefore, predicting the concentration of diacetyl is of particular interest for this research. As diacetyl is a flavour compound, its metabolism pathway is quite different from major components like the fermentable sugars. A separate model was adopted to represent the changes in diacetyl concentration.

This chapter overviews the development of a production model for the beer fer-

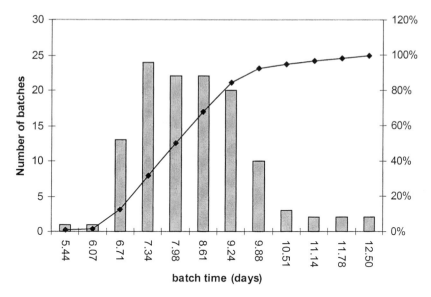

Fig. 25.2 Distribution diagram for the historical fermentation lengths of beer A.

mentation process. Further details have been published elsewhere of the sugar-yeast model[1] and of the diacetyl model[2].

25.3 Sugar-yeast model

A set of ordinary differential equations which describe the rate of fermentable sugar consumption and yeast growth was established as the core of the model. In a manner similar to that of Engasser[3], the rate of substrate consumption is modelled using the Monod equation:

$$\frac{d[\text{sugar}]}{dt} = -\frac{\mu_{\text{sugar}}[\text{sugar}]}{K_{\text{sugar}} + [\text{sugar}]}[x]$$

To account for the inhibition effect, inhibition coefficients k_i, are introduced. The original Monod equation is thus modified to the form

$$\frac{d[\text{sugar}]}{dt} = -\left(\prod_{i=\text{lower\_sugars}} \frac{k_i'}{k_i' + [i]}\right)\frac{\mu_{\text{sugar}}[\text{sugar}]}{K_{\text{sugar}} + [\text{sugar}]}[x]$$

The complete set of sugar equations is

$$\frac{d[F]}{dt} = -\overline{\mu_F x} = -\left(\frac{\mu_F[F]}{K_F + [F]} - \mu_{\text{inv}}[S]\right)x$$

$$\frac{d[G]}{dt} = -\overline{\mu_G x} = -\left(\frac{\mu_G[G]}{K_G + [G]} - \mu_{\text{inv}}[S]\right)x$$

$$\frac{d[S]}{dt} = -\overline{\mu_S x} = -(2\mu_{inv}[S])x$$

$$\frac{d[M]}{dt} = -\overline{\mu_M x} = -\left(\frac{k'_F}{k'_F + [F]}\frac{k'_G}{k'_G + [G]}\frac{\mu_M[M]}{K_M + [M]}\right)x$$

$$\frac{d[N]}{dt} = -\overline{\mu_N x} = -\left(\frac{k'_F}{k'_F + [F]}\frac{k'_G}{k'_G + [G]}\frac{k'_M}{k'_M + [M]}\frac{\mu_N[N]}{K_N + [N]}\right)x$$

where F, G, S, M, N, and x represent fructose, glucose, sucrose, maltose and maltotriose, and yeast content, respectively

The utilisation rate of fermentable sugars is dependent on the fermenter temperature. As indicated by Gee and Ramirez[4], the temperature dependence of the fermentation parameters follows the Arrhenius equation:

$$\theta = \theta_0 \exp(-E_a/RT)$$

where E_a is the activation energy, T is the temperature in K and R is the universal gas constant $= 8.314 \, \mathrm{J\,K^{-1}\,mol^{-1}}$.

Although the yeast cells multiply and grow as the fermentation proceeds, the growth rate is not constant. As an approximation, the rate of yeast growth can be represented as follows:

$$\frac{dx}{dt} = \mu_x x$$

where

$$\mu_x = \prod_{i=F,G,S,M,N} Y_{xi} \overline{\mu_i}$$

Fermentation data from five different batch fermentations of beer A were chosen for the estimation of model parameters. The parameters were estimated using the Solver^{TM} analysis extension in MS Excel^{TM}. The differential equations were solved by the Euler method. The squared sum of errors between the experimental data and the model predictions was minimised using a quasi-Newton method. Generally, the model fits well to the actual data. A typical model fit is shown in Fig. 25.3. To protect the intellectual property of the Lion Brewery, the original concentration (g/l), (ppm), specific gravity and temperature are here replaced by an internal measurement system having units of LBCU, LCU, LSGU and LTPU, respectively. The model was then validated with historical data from a large number of fermentations. Figure 25.4 compares the model predictions with the historical data trend for the changes in specific gravity during the course of the fermentation. The model predictions match the experimental data remarkably well. For most of the data point pairs, the differences between the predictions and the actual data are within one standard deviation.

The generic modelling capability of the sugar–yeast model was further demonstrated as follows. The specific gravity profile of another Lion Brewery product, beer B, was simulated, using exactly the same fermentation model as for beer A. Compared with beer A, beer B is fermented at a much higher temperature (Fig. 25.5(a)). The simulated

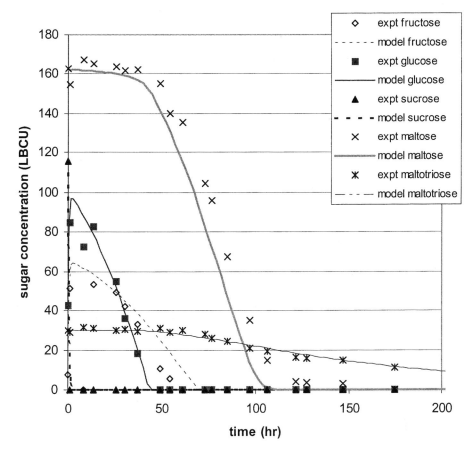

Fig. 25.3 Model fit for a typical set of beer A fermentation data (FV31350).

specific gravity profile over six sets of beer B data, FV31214, 31223, 31247, 31256, 31264, 31286 are shown in Fig. 25.5(b). With the exception of one set of data (FV 31274), the simulated profile compares extremely well with the actual production data.

25.4 Diacetyl model

The diacetyl profile was modelled using two different approaches: differential equation-based models and neural network-based models. The diacetyl model based on ordinary differential equations is described first.

The studies of Garcia[5], Gee[6], Inoue[7] and Rice[8] adopted differential equations as the basis for their diacetyl models. However, the equation sets that they chose were inconsistent. While Gee assumed that the diacetyl formation mechanism was dependent on the rate of yeast growth, Garcia suggested that the formation was related only to the rate of sugar and valine consumption. Gee, Inoue and Rice proposed that the decomposition of diacetyl should be spontaneous and follow simple

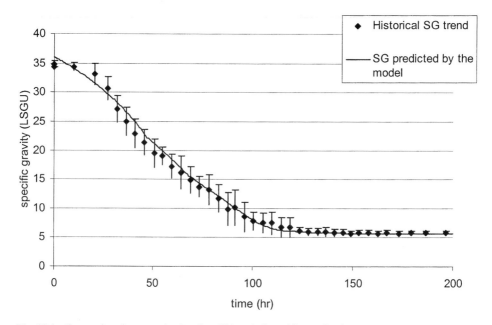

Fig. 25.4   Comparison between simulated and historical specific gravity data.

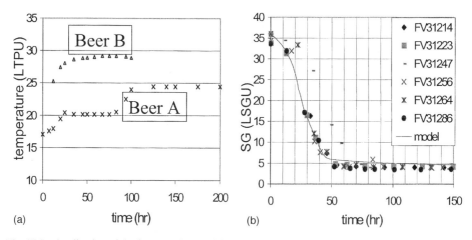

Fig. 25.5   Application of the fermentation model to another type of beer: (a) temperature profile and (b) specific gravity profile.

first-order kinetics, whereas Garcia adopted a time-varying coefficient for that reaction.

Since the exact diacetyl metabolism profile is not yet known, a number of possible model structures were explored. For the diacetyl decomposition part, all three models are identical. Differential equation models for diacetyl can be divided into two parts:

formation kinetics and decomposition kinetics. For the diacetyl formation part, model A emphasises the role of yeast growth, model B emphasises the specific gravity whereas model C is a combination of models A and B.

$$\text{model A}: \frac{d[\text{diacetyl}]}{dt} = k_f \frac{dx}{dt} x - k_{d0} \exp \frac{-E_a(k_d)}{RT} [\text{diacetyl}] x$$

$$\text{model B}: \frac{d[\text{diacetyl}]}{dt} = k_f \left| \frac{dSG}{dt} \right|^n x - k_{d0} \exp \frac{-E_a(k_d)}{RT} [\text{diacetyl}] x$$

$$\text{model C}: \frac{d[\text{diacetyl}]}{dt} = k_{f1} \frac{dx}{dt} x + k_{f2} \left| \frac{dSG}{dt} \right| x - k_{d0} \exp \frac{-E_a(k_d)}{RT} [\text{diacetyl}] x$$

Here, $E_a(k_d)$ is the activation energy for the decomposition of diacetyl.

Model A was taken directly from the study of Gee[6]. It assumes that the formation of diacetyl is proportional only to the rate of yeast growth: in other words, the diacetyl formation should cease when the yeast growth ceases. As illustrated in Fig. 25.6, the model does not really fit the experimental data. The experimental diacetyl peaks appear much later than the predicted ones. Inoue[9] observed this phenomenon when beer was fermented in large cylindrical fermenters.

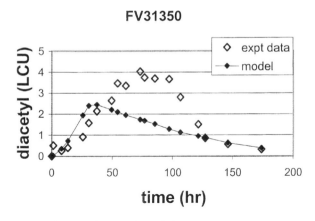

Fig. 25.6   Model A when fitted to the experimental data.

Model B assumes that the formation of diacetyl is linked to the rate of sugar consumption. Since total sugar content can be expressed as a linear function of the more readily available specific gravity measurements, the total rate of sugar consumption is thus replaced. The model does not match to that of the experimental data either (Fig. 25.7).

Model C is a combination of the first 2 models. As indicated in Fig. 25.8, it fits reasonably well to the experimental data. However, it tends to underestimate the diacetyl concentration near the end of the fermentation. For example, the experimental diacetyl contents for FV31350 were 0.543 LCU and 0.342 LCU at 160 h and 188 h whereas the Model C predictions were 0.342 LCU and 0.109 LCU, respectively.

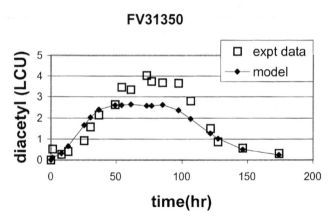

Fig. 25.7　Model B when fitted to the experimental data.

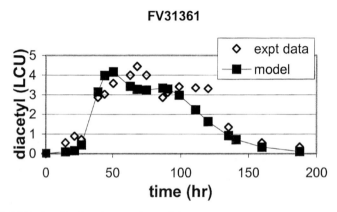

Fig. 25.8　Model C when fitted to the experimental data.

The architecture of an artificial neural network is inspired by the mammalian biological nervous system. A neuron is the basic building block for the artificial neural network. The connections between the individual neurons are expressed as weights. As in nature, the functionality of the network is determined largely by the connections between the basic elements[10]. Since an exact knowledge of the relationship between the variables is not required when establishing a neural network model, it should be easier to set up when compared with a differential equation-based model. Thus, a neural network model was adopted as an alternative way to model the concentration of diacetyl.

In neural net modelling, there are a number of choices to be made about the structure of the model. The structure of the neural network model adopted is shown in Fig. 25.9. A 2-4-1 three-layer (2 neurons in the first hidden layer, 4 neurons in the second hidden layer and 1 neuron in the output layer) back-propagation network (BPN) model with logistic sigmoidal neurons was selected. The neural network was trained using the Levenberg–Marquardt learning algorithm available in the

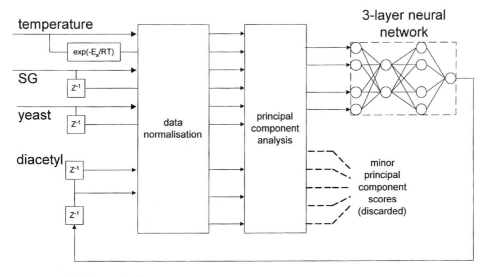

note: $z^{-1}y(t) = y(t-1)$

Fig. 25.9 Structure of the neural network-based diacetyl model.

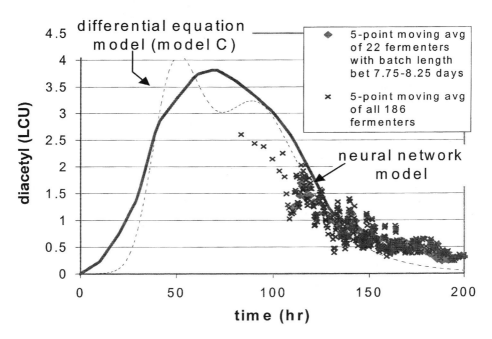

Fig. 25.10 The diacetyl models compared.

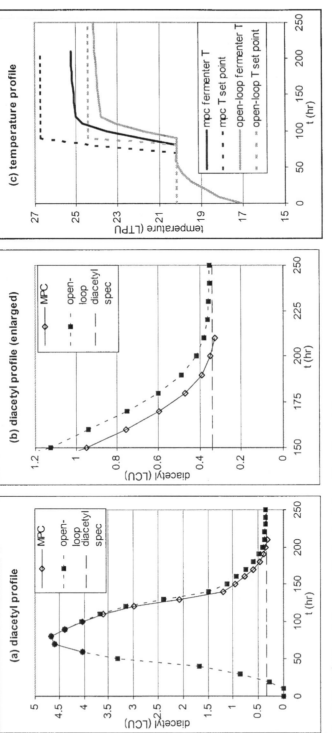

Fig. 25.11 Model predictive control in the simulation of beer fermentation.

Matlab^{TM} Neural Network Toolbox. The present and previous values of specific gravity, yeast and temperature and the previous values of diacetyl concentrations are taken as the inputs for the neural network. These input values are normalised and decorrelated (using the principal component analysis method) before being fed to the 3-layer neural network.

After training, the neural network model offers better predictions than does the differential equation method (Fig. 25.10). Its predictions match closely to the historical diacetyl data, especially near the finish of the fermentation.

25.5 Implications

After establishing both parts of the fermentation model, the possibility of using model predictive control (MPC) was evaluated by process simulations (Fig. 25.11). In the presence of unmodelled process disturbances, the model predictive controlled system works much better than the original open loop controlled system. The performance of MPC is still comparable with the original control scheme even when measurement noise is large. Production plant trials will be the next stage to verify the simulation results.

25.6 Conclusions

A fermentation model has been established for the production of beer A. The concentrations of the major fermentable sugars, diacetyl and yeast are included in this model. The behaviour of the major components, sugar concentrations and yeast, are modelled using differential equations. This sugar–yeast system has a good generic modelling capacity. It has been used successfully to predict certain production characteristics of another Lion Brewery beer. For modelling diacetyl behaviour, the structure available in a neural network model provided a better representation of observed process behaviour than the best available differential equation model.

Acknowledgements

This work was supported by Technology New Zealand through the GRIF scheme (Contract Number: LIO801). The financial support and technical assistance of Lion Nathan (Lion Brewery, Auckland) is also gratefully acknowledged.

References

(1) Yuen, D.C.K. and Austin, P.C. (2000) Industrial beer production by *Saccharomyces cerevisiae*: a kinetic model and its applications, *J. Biochemical Engineering*, in press.
(2) Austin, P.C. and Yuen, D.C.K. (2000) An investigation of the benefits of using a model predictive control system in beer fermentation, *J. Process Control*, in press.
(3) Engasser, J.M. *et al.* (1981) Kinetic modelling of beer fermentation in a European Brewery, *Proceedings of the 18th European Congress*.

(4)   Gee, D.A. and Ramirez, W.F. (1988) Optimal temperature control for batch beer fermentation, *Biotechnol. Bioeng.* **31**, 224–234.

(5)   Garcia, A.I., Garcia, L.A. and Diaz, M. (1994) Modelling of diacetyl production during beer fermentation. *J. Inst. Brew.* **100**, 179–183.

(6)   Gee, D.A. and Ramirez, W.F. (1994) A flavour model for beer fermentation. *J. Inst. Brew.* **100**, 321–329.

(7)   Inoue, T. (1977) Forecasting the vicinal diketone content of finished beer. *J. Amer. Soc. Brew. Chem.* **36**, 9–12.

(8)   Rice, J.F. and Helbert, J.R. (1973) The kinetics of diacetyl formation and assimilation during fermentation. In *Proceedings of the American Society of Brewing Chemists' Annual Meeting*, New Orleans, ASBC.

(9)   Inoue, T. (1992) A review of diacetyl control technology, *Technical Report of Kirin* **33**, 91–94.

(10) Demuth, H. and Beale, M. (1998) Neural Network Toolbox User's Guide (Version 3.0), Mathworks Inc., Natick.

26 Yeast Management and Fermentation Performance: a Brewer's Perspective

WARREN QUILLIAM, GAVIN HULSE and
ANNA CAMERON-CLARKE

Abstract The modern-day brewer is challenged with achieving very tight targets of productivity, throughput and plant efficiency. In the light of this, it has become increasingly more difficult for the brewer to provide the yeast with the stress-free environment that it requires in order to produce the desired level of end-product quality. More often than not, the brewer exacerbates the stress environment, in an attempt to achieve these targets, by forcing the yeast to perform, and in so doing, ending up in a spiral of poor performing yeast and declining product quality. Practices such as over-pitching, increasing collection and fermentation temperatures, and keeping the yeast in contact with the product for prolonged periods of time in order to achieve attenuation, all contribute to the development of 'off-flavours' associated with yeast stress. The design of equipment used for yeast recovery and subsequent re-pitching, as well as the practices in this part of the operation also influence the performance of yeast in subsequent fermentations. This chapter examines yeast handling in typical modern breweries and the ramifications that it has on final product quality.

26.1 Introduction

Like many other businesses in today's environment, the business of brewing beer is continuously under pressure to improve productivity, make better use of capital equipment, put the squeeze on raw materials, all in an effort to survive in an environment of declining sales and increasing competition.

The challenge facing the modern day brewer is therefore to be work-smart, be more efficient, eliminate waste, produce more from less, and of course achieve the expected quality standards while doing all of this! In addition, the brewer is expected to achieve these quality standards in a consistent manner under ever changing conditions which more often than not will lead to corrective action in conflict with good brewing practice. Unfortunately these actions can very quickly result in a spiralling decline in quality, and eventual consumer reaction.

While brewers have differing opinions as to which part of the brewing process 'makes or breaks' the product, it is generally accepted that no matter how good the raw materials, or how appropriate the equipment in use is, if the yeast is not happy (perhaps even deteriorated), it is highly unlikely the expected quality standards in the final product will be achieved.

26.2 Expectations from the yeast

The modern day fermentation process inherently imposes a stressed environment on the yeast by virtue of a number of factors, the most important being as follows.

26.2.1   High gravity

Many brewers in an attempt to achieve improved productivity are brewing 16, 18 and 20°Plato beers quite successfully, but normally at the cost of quality in the form of increased acetaldehyde and sulphur presence. If the brewer is happy to limit the number of generations recovered to one or two re-pitchings, then this is not normally an issue. The temptation to recover and re-pitch the yeast for up to eight or more generations is ever present, despite the possibility of flavour-negative compounds.

26.2.2   High levels of alcohol

Associated with high gravity brewing is the fact that yeast, particularly in the propagation stages, can become exposed to high levels of alcohol for prolonged periods of time through delays in transferring the yeast into the brewery propagator. This problem is particularly evident in breweries that distribute yeast propagations from a centralised facility where considerable distances have to be covered and transport delays are routine. If the propagation is carried out in high gravity wort, prolonged exposure of this fresh yeast to levels of alcohol in excess of 7% abv cannot be expected to do the yeast much good, especially if it is to be used for the next seven generations.

26.2.3   Fermenter design

While the trend towards large multi-brew fermentations has revolutionised beer production and certainly helped with keeping the price of beer within the consumer's reach, it has brought with it the potential of additional stress to the yeast in the form of increased static head. Instead of contending with a static head of 2 or 3 metres in the fermenters of old, the yeast cell is faced with a static head of 20-odd metres in these large multi-brew fermenters. This problem can be further compounded by the multi-filling mechanism, potential layering in the vessel, and problems with CO_2 toxicity.

26.2.4   Thermal shock

The use of very low refrigerant temperature to control fermentation, and also to cool recovered yeast, can have dire consequences for yeast in poor condition and susceptible to autolysis through sudden changes in temperature. Yeast recovered from the cone of the fermenter has been shown to reach temperatures of up to 20°C, and this can be reduced to temperatures as low as 2°C through the yeast recovery chiller within a matter of seconds. This yeast would have to be exceptionally healthy in order not to sustain any damage.

26.2.5   Process delays

Delays in the recovery of the yeast from the fermenter cone once the fermentation has run its course not only negatively affects the product in the fermenter, but also will stress the yeast for future fermentations as a result of temperature build-up in the cone and shortage of nutrient. Delays to the pitching of yeast, despite the traditional low temperature storage conditions, because of the lack of nutrients will once again stress the yeast, with obvious consequences to subsequent fermentation performance.

26.2.6 *Propagation top-up procedures*

Timing of the top-up stages of the propagation process has been shown to affect the performance of yeast radically. If this process is too early, then the brewer runs the risk of not allowing the yeast to acclimatise to the more complex sugars, and as such will battle in the transition from glucose to maltose and maltotriose. If the top-up process is too late, then the yeast cells progress into the stationary phase and subsequent stress condition, which once again negatively affects performance in subsequent fermentations. See Fig. 26.1.

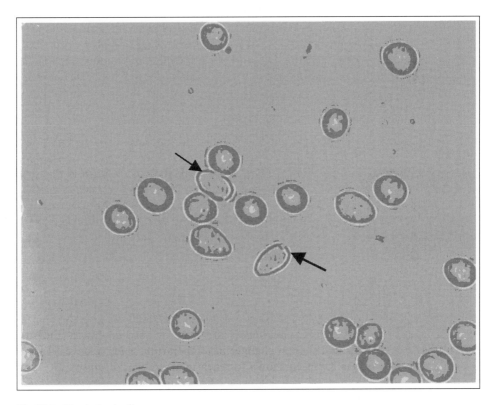

Fig. 26.1 Yeast ghost cells.

26.2.7 *Mechanical damage*

Large, highly automated plants with their associated long distances of pipework, automated valves, booster pumps and sometimes high speed centrifuges, have the potential to cause additional stress to the yeast which we expect to perform at its peak. While many references have shown that yeast is normally relatively resistant to mechanical damage, if it is in poor condition resulting from any form of stress, then it will most certainly not be in a position to withstand any form of mechanical damage, as Table 26.1 shows. Inappropriate automation, especially relating to yeast cropping, coupled with the inability to visually monitor the cropping process in highly modernised plants, does not help the process of stress minimisation either.

Table 26.1 Effect of mechanical damage on yeast protease release where the yeast is in poor condition.

Sample	Protease results
FV 76	0.044
After 3000 kg FV 76	0.070
After 3000 kg YCV 2	0.103
After 6000 kg FV 76	0.106
After 6000 kg YCV 2	0.109
After 9000 kg FV 76	0.088
After 9000 kg YCV 2	0.131
End of Crop	0.072
After 15′ mixing	0.124
After 30′ mixing	0.187
After 45′ mixing	0.225
After 60′ mixing	0.308
After 90′ mixing	0.613
After 120′ mixing	0.294
Batching out	0.688

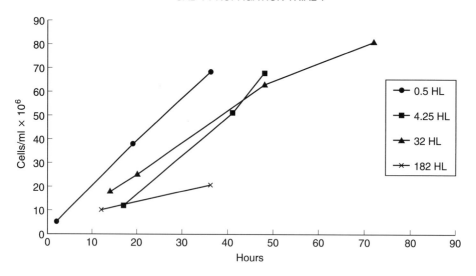

Fig. 26.2 Yeast cell count versus propagation time, showing the effect of delayed 'top-up' (31 hl stage) on cell growth (182 hl stage).

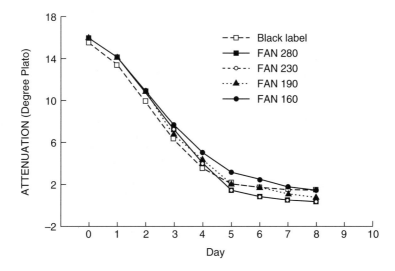

Fig. 26.3   Attenuation rates with different FAN levels using first generation yeast.

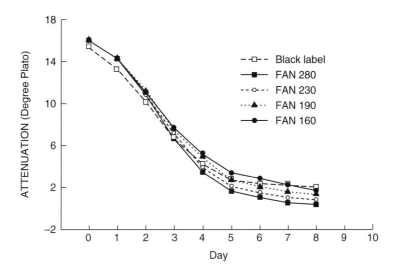

Fig. 26.4   Attenuation rates with different FAN levels using sixth generation yeast.

So during 'normal' operation, the yeast is already expected to perform under difficult and variable stress conditions. It is important, therefore, under these conditions that the vitality level of the yeast be kept at its highest possible level, where it is still able to perform under conditions of stress. With the expectation to maximise production efficiencies by working more smartly and achieving more from less, the brewer tends to be forced into practices which are less than desirable in the interest of producing a flavour stable product.

An example of this would be a fermentation carried out with young yeast that had been subjected to some form of stress. Ignoring for the time being the flavour implications, the typical situation would be a so called 'tailing' fermentation where the difference between the present extract and limiting extract is higher than planned. If this yeast is of a young generation, it would be probably be used for seven to ten more fermentations. So very soon the fermentation cellar would be filled with product that has not fully attenuated, and an extract loss account amounting to thousands of pounds per week. It is at this stage that panic sets in, resulting in 'quick fixes' to attempt a resolution of the problem.

The first attempt at resolving this problem is to speed up the fermentation rate. Normally this is achieved by collecting the wort at higher temperatures, fermenting the product at higher than normal temperatures, increasing the level of oxygenation and increasing the pitching rate. Normally this is also associated with a delay in the recovery of yeast in the hope the remaining fermentable sugars are taken up prior to chillback.

While these actions may help alleviate the problem in the short term, in the long term, the yeast falls into a spiral of declining quality as a result of the additional stresses it sustains.

26.3   Product quality and flavour implications

26.3.1   *pH*

With increased autolysis comes an increase in pH, and while this may assist with increased colloidal stability, it can detract from product drinkability and expected brand profile. Monitoring pH changes of a product which has attenuated but for which the crop has not yet been recovered can reveal a pH rise of 0.3 to 0.4 units. The increase in pH can also be detected in the recovered yeast slurry which increased by in excess of a whole unit above the product pH.

26.3.2   *Protease activity*

With the presence of foam on the final product considered as extremely important by most brewers and beer consumers alike, an increase in protease activity from yeast autolysis heralds a decrease in the ability of the product to support a stable foam. With protease levels of 0.01 (ΔA 574 Nm) in yeast slurries and 0.001 (ΔA 574 Nm) in bright beer regarded as maximum acceptable levels, autolysis resulting from stressed yeast pushes these levels up 10-fold with associated decreases in foam results of 50 to 100 Nibem foam seconds (see Fig. 26.5).

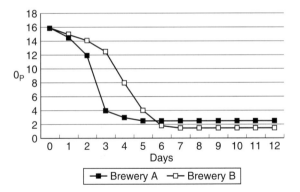

Fig. 26.5 Typical fermentation charts showing the effects of fermentation rates on final attentuation.

26.3.3 *Acetaldehyde*

Fermentations carried out with stressed yeast showed increases in acetaldehyde results to levels more than double the expected. With the perception that the lower the acetaldehyde levels in the final product, the better the drinkability, minimising the acetaldehyde production to levels of less than 3 ppm is not achievable using stressed yeast, which tends to produce levels in the final product in excess of 10 ppm (see Fig. 26.6).

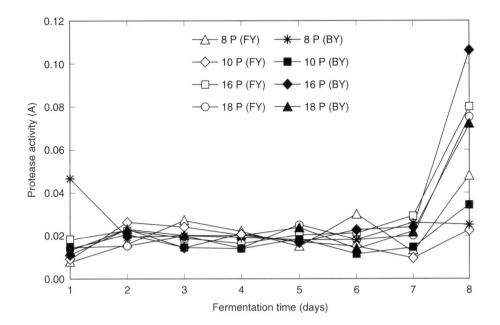

Fig. 26.6 Effect of delayed cropping on yeast protease release (protease activity in 8, 10, 16 and 18°Plato worts inoculated with laboratory propagated yeast (FY) and brewery yeast (BY)).

26.3.4 *Sulphurs*

An increase in sulphur dioxide levels as a result of autolysis may be regarded as being positive by virtue of its anti-oxidant properties (see Fig. 26.7). Unfortunately this is not the only sulphur compound that is associated with autolysis, and associated increases in hydrogen sulphide and methanethiol lead to detracted drinkability.

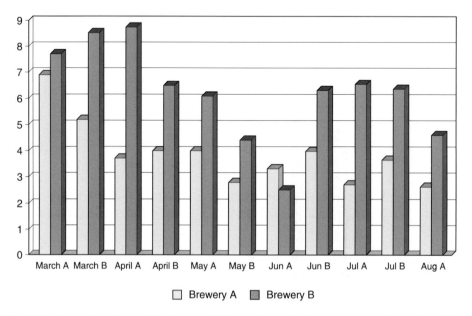

Fig. 26.7 Comparison of acetaldehyde levels in final products from brewery A (yeast in good condition) and from brewery B (yeast in poor condition).

26.3.5 *Haze*

The release of cell contents into the supporting medium has been shown to create filtration problems, in many cases producing an unfilterable haze which carries through into the final product.

26.3.6 *Fobbing*

The increased fermentation rate can also have an impact on a number of issues. As a result of the fermentation vigour the vessel tends to fob, resulting in beer loss, loss of bittering substances and loss of foaming potential, with most of the foam protein disappearing down the drain or into the carbon dioxide recovery plant. The end result of this is highly variable bitterness results, poor in-package foams, and carbon dioxide purification problems by virtue of the fact that the carbon dioxide plant has been overloaded with organic material from the fobbing fermentations.

26.3.7   *Taste*

A fast fermentation (greater than 6°Plato/24 h) can result in extended FAN reduction and low pH, producing a product that is harsh drinking and bordering on being sour. With associated autolysis, these harsh flavours are accentuated, resulting in a beer which the consumer may not rush back to the bar counter to reorder. Using yeast which has been subjected to stress can significantly alter the brand profile.

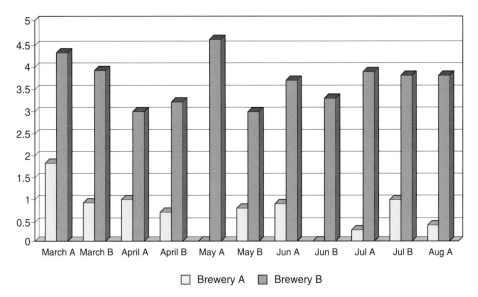

Fig. 26.8   Comparison of SO_2 levels in final products from brewery A (yeast in good condition) and brewery B (yeast in poor condition).

26.4   Are we therefore achieving more for less?

As with most brewing issues, being a successful brewer means having the ability to manage compromises. By speeding up fermentation rates in order to alleviate fermentation problems, the brewer is in actual fact creating a host of other more serious problems, not only relating to product quality but also to product cost, largely through poor attenuation.

26.5   Finding the balance

What is the answer? What does the brewer have to do in order to achieve cost and quality efficiency? Short of finding a strain of yeast which is invincible and totally able to withstand all forms of stress thrown at it, the brewer has to ensure that the yeast stresses are minimised at all stages of the process from propagation to scrapping. In

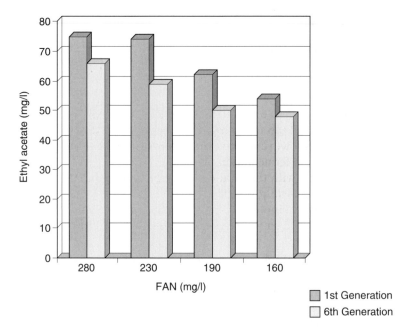

Fig. 26.9   Effects of yeast generation on ethyl acetate at different levels of FAN.

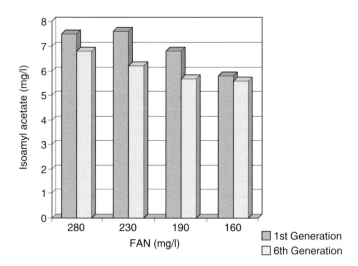

Fig. 26.10   Effects of yeast generation on isoamyl acetate at different levels of FAN.

essence, the yeast has to be nursed though the process, possibly by a dedicated brewer who takes full responsibility for the well-being and performance of the yeast. The following twelve point plan should provide the desired end result for yeast handling and subsequent fermentation performance.

26.6 The twelve point plan

(1) Ensure that the propagation technique is such as to produce biomass rather than alcohol, and then minimise the contact between the propagated yeast and the alcohol.

(2) Top up the various propagation stages at the appropriate times. Each brewery would need to establish their specific exponential growth pattern to determine the ideal top up timing. This should occur towards the middle of the exponential growth curve to expose the yeast to the sugar spectrum it can expect in the main fermentation process.

(3) Provide the yeast with the ideal oxygenation regime during propagation to ensure optimal growth potential.

(4) Provide the fermentation with enough nutrients (especially FAN) from the raw materials (paying more for better quality malt invariably costs far less at the end of the day).

(5) Provide the yeast with stress-free conditions during the fermentation process. This relates to temperature control and rate of oxygenation in particular. The wort should be collected at a temperature suitable for ensuring a slow start to the fermentation, and then rise to a fermentation temperature suitable for the brand profile. Sufficient oxygen must be provided to enable yeast development but also accommodate the required brand profile characteristics. Starving the yeast of oxygen might give the product a high level of ester but will also cause the development of stress products.

(6) Minimise temperature shocks. If a refrigerant temperature of close to zero degrees Celsius is capable of controlling the fermentation process, it would certainly be of benefit to the growing yeast cells, which would not be subjected to a temperature shock if the refrigerant were at a lower temperature. Similarly the cone temperature, which is often controlled at -4 degrees Celsius, could be contributing significantly to yeast autolysis at the interface. Consideration should also be given to a two-phase water chiller for reducing the temperature of recovered yeast, rather than the sudden reduction of temperature using propylene glycol or compressed ammonia at very low temperatures. The rate of chillback of the fermentation should also be prolonged to ensure that the possibility of temperature shock is minimised.

(7) Recover the yeast on time. When the fermentation has reached its attenuation limit there is no need to keep the yeast in contact with the product. Yeast which is still in suspension will mop up any residual fermentable sugars that may remain in the beer. More harm than good will result from delaying yeast recovery. It is also essential that all the yeast is removed and that slurry consistencies and crop sizes are monitored as indicators of fermentation performance.

(8) Scrap the 'tired' yeast. Phenomenal improvement in yeast performance has been reported if the first 10 to 15% of the recovered yeast is scrapped. Improvements in protease levels, attenuation, pH and flavour were reported after removal of the early flocculating 'tired' yeast from the base of the fermenter prior to recovering the crop.

(9) Ensure homogeneity, gas stripping and temperature control in recovered yeast. Gentle but effective agitation is critical to minimising yeast stress through the achievement of homogeneity and the removal of entrapped carbon dioxide. It is also critical that the temperature increase of the recovered yeast is restricted to a few degrees Celsius prior to pitching, or the quality of the subsequent fermentations will be negatively affected.

(10) Minimise the recover to re-pitch time. It is not always practical to minimise the occupancy of the recovery and pitching vessels, particularly if the brewery uses several strains of yeast. The longer the yeast spends away from nutrient, despite the temperature, the more prominent the stress placed on that yeast.

(11) Minimise mechanical damage. Consideration has to be given to the gentle handling of the product. Care needs to be given to the selection of pumps, the complexity of the pipework, the number of valves, the action of agitators and the design of chillers in order to protect the yeast from unnecessary mechanical stress.

(12) Use the best yeast … scrap the rest. With every fermentation potentially different to the others, it is critical that the brewer responsible for selecting the yeast scrutinises every aspect of the parent fermentation, ensuring that the best yeast is selected for the subsequent fermentations.

26.7   Conclusions

Yeast that is handled correctly can produce high quality product in a cost effective manner, despite the stresses that are clearly evident in modern brewery environments. Achievement of this output does, however, require the undivided attention of the brewer, who needs to monitor not only yeast performance but all aspects of the production process that can influence this performance.

Acknowledgements

I would like to thank Mr. Gavin Hulse and Mrs. Anna Cameron-Clarke for their efforts in generating data in support of the views expressed in this chapter.

Index